D0881172

Understanding the Mathematics Teacher

A Study of Practice in First Schools

Understanding the Mathematics Teacher

A Study of Practice in First Schools

Charles Desforges

and

Anne Cockburn

The Falmer Press

(A member of the Taylor & Francis Group)

London New York and Philadelphia

UK The Falmer Press, Falmer House, Barcombe, Lewes, East Sussex, BN8 5DL

USA The Falmer Press, Taylor & Francis Inc., 242 Cherry Street, Philadelphia, PA 19106-1906

First published 1987

Library of Congress Cataloging-in-Publication Data

Desforges, Charles.
Understanding the mathematics teacher.

Bibliography: p.
Includes index.
1. Mathematics—Study and teaching (Elementary)—Great Britain. 2. Mathematics teachers—Great Britain. I. Cockburn, Anne. II. Title.
QA135.5.D475 1987 372.7'0941 87–15593
ISBN 1–85066–212–6
ISBN 1–85066–213–4 (pbk.)

Jacket design by Caroline Archer

Typeset in 11/13 Bembo by
Imago Publishing Ltd, Thame, Oxon

Printed & bound in Great Britain by
Redwood Burn Limited, Trowbridge, Wiltshire.

Contents

Acknowledgements

We have had a great deal of help in conducting this research. It would not have been possible without the financial backing given over two years by the University of East Anglia. Hugh Sockett carried our bid for support through the intense competition for scarce University funds. We are indebted to him for his midwifery.

Very large numbers of friends, associates and students have discussed our ideas with us as they have evolved from first thoughts to final manuscripts. Outstanding amongst these have been Pat Brittan, Anne Desforges, Carolyn Horne and Deirdre Pettitt. More than 1000 teachers on in-service courses up and down the country have helped us evaluate our data and refine our interpretations. Several advisers and teachers have read drafts or part-drafts of the manuscript and given us good advice. We are very grateful to all these people. Our work has been enormously enriched by their thoughts. Perversely, perhaps, we have not always taken their advice in the way intended. In consequence, errors of judgment in the following pages are all ours.

We would like to thank Sonia Mason for typing the manuscript from our scrawl, for coping with endless alterations, for not dreading the references too much and for keeping us to schedule in a way that brooked no argument.

Most of all we are indebted to the teachers who are the focus of this research. Because serious teachers invest so much of themselves in their classrooms, it takes a special kind of professional to lay themselves open to the analysis of outsiders. The teachers whose practices are discussed in this book gave unstintingly to the task of professional evaluation. Each started as a collaborating colleague and each became a friend. We hope our version of their labour shows them how much they taught us. We know they would value our learning as much as our thanks.

Charles Desforges
and Anne Cockburn
April 1987

Chapter 1

The Trouble with Mathematics Teaching

Introduction

This book is about some of the problems facing teachers as they try to develop young children's mathematical competence. In it are reported the results and analyses of a two-year in-depth study of the practice of seven experienced first school teachers.

The study was provoked by our suspicion of the bad press generally received by teachers of mathematics at all levels. It seems to have been assumed by teachers' critics that the only barriers between pupils and a mathematical wonderland of experience are those erected by teachers' conservative attitudes and their poorly informed practices.

Clearly there is a problem in mathematics education. There is a huge gap between aspiration and achievement. Strenuous efforts over decades seem to have left the problem untouched. A great deal of work has been done on understanding children's mathematical thinking, on identifying and sequencing educational objectives and on designing attractive teaching materials. The work has generated plenty of advice for teachers most of which does not seem to be practised in most classrooms. Seen in this light it is easy to appreciate why teachers are the butt of criticism. Who or what else could be to blame?

Whilst this conclusion looks reasonable, we have worked with too many teachers to find it convincing. To blame teachers is to assume that once one understands children's thinking, holds clear objectives and has copious supplies of attractive teaching materials to hand, then everything is possible. In our experience as teachers and researchers, nothing could be further from the truth. We have worked with many teachers who were well informed on all these matters and yet who routinely failed to meet their own aspirations which they shared with mathematics educators. We do not believe that this failure has been due to idle or uninformed teaching. On the contrary, we suspect that the failure lies in the unwillingness of the mathematics education establishment to take seriously the complexities of the teachers' job.

Of course the problem of mathematics education is a many headed monster. It certainly involves an appreciation of the psychology of learning, the structure of mathematics, the demands of society and the design of curriculum materials. But experience suggests that it is also a classroom problem. Teachers work in the face of a number of complex and interacting classroom processes. These forces are manifest in the problems of communication, cooperation, management and evaluation. Experienced teachers know this. Successful teachers adapt to these processes. Nonetheless the processes are not well understood. Adaptation appears to be at an intuitive or pragmatic level. Recent American research has begun to reveal just how necessary it is to the teacher to adapt to these forces and yet how costly — in educational terms — the accommodations might be.

In the light of this research, and in view of our scepticism about other explanations of the limited success of mathematics education, it seemed to us essential to understand how classroom processes impinge on teachers as they endeavour to teach mathematics. If our suspicions about the force of these processes were correct, designing curriculums in the absence of this understanding is like designing flying machines on the principles of aerodynamics but ignoring gravity. Such machines would look beautiful in the air if only they could be got off the ground.

In the rest of this chapter we develop the professional and theoretical rationale behind our research. In subsequent chapters we report our findings. We show that the teachers' task is infinitely more complex than that assumed in the literature on mathematics education. In view of our account of this complexity we argue that current approaches to the improvement of mathematics education will have little effect. We suggest that those who provide — both materially and conceptually — for mathematics teaching will have to change their ways dramatically if teachers are to be supplied with a context in which to change their practices.

What Are the Shortcomings of Mathematics Education?

There seems to be little point in repeating here the very extensive surveys of the state of mathematics teaching and learning. Exhaustive commentaries are readily available (Cockcroft, 1982; Romberg and Carpenter, 1986; Suydam and Osborne, 1977; Weiss, 1978). Suffice it to note here that whilst some lurid claims have been made about wholesale failure on the part of mathematics teachers, detailed research indicates more specific and contained deficits.

It has been claimed, for example, that after ten years of schooling involving more than 1000 hours of mathematics instruction, school leavers find it hard to do simple calculations (ACACE, 1981; Trafton, 1979), are mathematically inarticulate and have little confidence in their ability (Easley and Easley, 1983), adults are prone to seize up when faced with mathematical problems (Buxton, 1981), and secondary school pupils have a poor understanding of number (APU, 1982; Ward, 1979). It should be noted that these claims refer to experience on both sides of the Atlantic.

These sweeping criticisms were not entirely substantiated by the Cockcroft report. After extensively canvassing employers, Cockcroft concluded that levels of attainment were not as bad as the Committee of Inquiry had been led to expect. Basic computational skills were not criticized. The anxiety about mathematics education seemed to be aroused more by the observation of children's limited capacity to use their skills appropriately to recognize, represent and solve problems. It seems that children do not need further doses of basic skills training. Rather, they appear to need to acquire a 'feel' for number and other mathematical processes, together with a degree of intellectual autonomy which would enable them to go beyond routine calculations and to solve real-life problems involving mathematical thinking. They need to learn to use their skills with flexibility. They need to learn to think with mathematics rather than merely respond with routines.

The concern about mathematics education is not new. Cockcroft (1982) showed that similar complaints can be traced back to the earliest HMI documents. Despite the antiquity of the anxiety about disappointing standards, mathematics educators remain convinced that schools can significantly improve on current performance.

Why Should We Try to Improve on Present Attainments?

It has been argued that modern citizens need to be confident with mathematical forms of reasoning, skills and knowledge in order to be able to survive and profit in their social world and to contribute properly to the development and civilized operation of a technological society. Since citizens appear to operate successfully alongside the evident persistence of mathematical failure this line of argument seems to lack cogency. It is, nonetheless, heartfelt and contributes to set the emotional tone of the debate. The tone is, apparently that we must all try harder. Unfortunately, emotional tone has never been conducive to clear thought.

There are, at first sight, more convincing arguments as to why we might feel disappointed with the products of mathematics teaching and why we might reasonably expect to do better.

There is a growing body of evidence which suggests that before children come to school they are inventive mathematical thinkers — albeit with a range of application necessarily constrained by their limited experience. There is evidence, for example, that pre-schoolers try to impose sense on the business of counting and operate counting as a rule-governed procedure (Gelman and Gallistel, 1978). Studies of young children's errors in counting have been interpreted as showing a persistent search for pattern. Three-year-olds appear to invent the notion of conservation of number for small quantities. They also invent rules for increasing, decreasing and comparing quantities (Gelman, 1977). They develop and operate simple mathematical processes (including addition and subtraction) for concrete objects (Hughes, 1986). Shown how to add up two small quantities by setting out both sets and counting the lot, they quickly — under conditions of practice — appear to invent short cuts including counting-on from the larger quantity (Resnick and Ford, 1981). They have a rich variety of strategies for dealing with remainders in problems involving sharing (Desforges and Desforges, 1980). Several recent studies have shown that young school-age children have strategies for solving problems (including word problems) which are more efficient and conceptually based than the mechanical procedures they are taught in school (Romberg and Carpenter, 1986). When these spontaneous achievements are compared with the results of schooling the latter are bound to be disappointing. Worse, they may be taken to be a drastic waste of natural resources. If young children are so naturally mathematically inventive and have, untaught, a capacity to impose mathematical structure on their experience, surely, it might be argued, it is not too much to expect that they should be better — and certainly not worse — than this as a consequence of schooling. In the light of studies of young children's thinking the effects of schooling might be seen not merely as a disappointment but as a disaster.

Such a conclusion is less plausible than it appears at first sight. It assumes that the natural thinking processes of a young child remain available to older children in a 'natural' world and that any decline in natural propensity must be attributable to the effects of schooling. This view has been seriously questioned (Bereiter and Scardamalia, 1977). Bereiter and Scardamalia have argued that inventiveness is driven by curiosity. Curiosity poses the problems which inventiveness must solve. They also point out that there is a drastic decline in curiosity in all species at the termination of infancy. They suggest that curiosity has survival value for the very young because they must learn rapidly in order to adapt to their environment. Curiosity however, carries a risk. The curious are vulnerable. Curiosity, in this view, becomes rapidly susceptible to the law of diminishing returns. Once a basic corpus of habits has been acquired, increments make very little difference and the risk to gain them is not worth taking. Bereiter and Scardamalia argue that just as curiosity has survival value for the very young, so the decline of curiosity has survival value for the relatively mature. Natural inventiveness will not operate in the absence of questions or problems consequent upon the decline of natural curiosity. In this view it is an error to blame a loss of inventiveness and curiosity on the effects of schooling. Indeed it is quite counter-productive to do so since, Bereiter and Scardamalia argue, it deludes teachers into feeling that they can — indeed they must — resuscitate and capitalize on something which has, in fact, diminished as a consequence of natural development and adaptation. This delusion, suggests Bereiter and Scardamalia, may be reinforced by the occasional positive outcomes in which, by dint of enormous effort, teachers do successfully — albeit temporarily — rekindle children's inventiveness. If this line of argument is taken to be plausible it is obviously foolish to set the expectations of schooling in the light of the somewhat ephemeral thinking characteristics of the young child. What we learn after the age of 7, argue Bereiter and Scardamalia, we must learn as a deliberate act rather than as a convenient but incidental spin-off from curiosity.

To raise questions about the relevance of studies of young children's spontaneous mathematical thinking to the setting of educational expectations is not to dismiss these studies out of hand; nor is it to argue that schools cannot improve on current achievements. The relevance of such studies to the design of educational processes is discussed in chapter 8. The point being made here is simply that it is naive to extrapolate from these studies on the untested assumptions that young children's inventive thinking processes continue to be spontaneously available to them as they grow to school age and that only school limits their operation.

The best evidence that schools can — and hence should — enhance the quality of mathematics teaching would come from studies which showed improvements to be consistently associated with the use of particular, practicable techniques of teaching. Studies which contrast the achievements of above average teachers with those of below average teachers are no use in this respect. The 'above average' simply become a stick with which to beat the rest. Such an argument makes a nonsense of the notion of average. It is not enough, in setting standards, to show that some teachers do better — and hence should be a guide to the definition of expectations. It is necessary to give some coherent account of the methods they use to do better and to show that these methods are transferable to other teachers.

Some such studies do exist and give lead to the reasonable expectation that schools can improve on mathematics teaching. The best documented studies in this respect have researched the use of the techniques of mastery learning (Bloom, 1976) or direct instruction (also known as 'systematic teaching' or 'explicit instruction' or 'structured teaching') (Good, 1979; Rosenshine and Stevens, 1986). The use of these techniques has been shown to be associated with learning gains by young children in mathematics which are significantly more than learning emanating from the use of other techniques commonly seen in classrooms.

The effective teacher using direct instruction:

> Begins the lesson with a short review of prerequisite learning ... announces the goals of the lesson ... presents new material in short steps ... gives students practice after each step ... gives clear and detailed explanations and instructions ... gives a high level of active and successful practice .... (*Ibid*, p 377)

In Rosenshine and Stevens view, '... all teachers use some of these behaviours some of the time, but the most effective teachers use most of them almost all of the time'. (p 377). Rosenshine and Stevens argue that the techniques are consistent with modern psychological theories of learning and remembering. New knowledge is linked to old knowledge, new material is actively processed, reviewed, rehearsed and summarized to aid its retention and only a little material is introduced at a time so that there is no overload of limited memory capacity. Their review of research suggests that the methods are particularly appropriate and successful with younger pupils and with, 'arithmetic facts ... processes and skills ... application of skills to new situations ... mathematical computation ... and algebraic equations' (p 378).

Despite the evidence that children whose teachers have used the techniques of direct instruction consistently do significantly better on mathematics tests than children whose teachers do not use these techniques systematically, such approaches to mathematics teaching are not ubiquitously well received by mathematics education experts, as we shall now see.

Common Practice, Obvious Problems and Obvious Solutions

Extensive surveys of classroom practice indicate that far from encouraging and developing young children's mathematical inventiveness, the vast majority of school teachers have children playing a passive-receptive role as learners. The teachers induct children into mathematical routines by — in the main — taking them through a commercial mathematics scheme. American studies (Fey, 1979; Sirotnik, 1983; Welch, 1978) show that the average mathematics lesson is dominated by the teacher. The teacher reviews previous work, explains and demonstrates new procedures and the explanation is followed by the children's carrying out routine pencil and paper practice exercises. Whilst the children practise, the teacher circulates amongst them to provide encouragement and correction. Welch (1978) observed that, 'The most noticeable thing about maths classes was the repetition of this routine' (p 6).

In our own study (Bennett *et al*, 1984) of infant classes in England, children worked less as a class and much more at their own rate. Work rate and class organization apart, the role as learner of the English child was identical to that of the American child. The teacher explained a mechanical procedure in the mathematics scheme and the children did pencil and paper exercises to practise the routine. Eighty per cent of mathematics work observed in our study conformed to this pattern. Although, on the evidence of this one study, English children get much more individual direction than their American counterparts, their role as learner and the nature of their mathematical experience are identical in principle. The children work mechanically to reproduce teacher performances in order to make progress through their mathematics workbooks.

The striking thing about this approach to teaching is that it bears a remarkable resemblance to Rosenshine and Stevens' model of effective instruction. Both in common practice, as revealed by survey research and in Rosenshine and Stevens' model of effective teaching derived from experimental studies, there is an emphasis on teacher review, teacher explanation and direction, children's practice and subsequent feedback from the teacher. Presumably where common practice fails, as Rosenshine and Stevens hinted, is in the persistence and clarity of particular performances.

But in the eyes of mathematics educators there is a more obvious and indeed fatal flaw in common traditional practice. In the words of Romberg and Carpenter (1986), 'Traditional instruction is based on a metaphor of production in which students are seen as "raw material" to be transformed by "skilled technicians". The view of learning and teaching upon which this metaphor is based is no longer appropriate.' (p 850).

Romberg and Carpenter (*ibid*) identify three serious limitations to the 'production' approach to mathematics teaching. First, the child meets formal mathematics as a set of concepts and skills divorced both from reality and from enquiry. The pencil and paper exercises lack a link with real problems and demand no skill other than reproduction for their solution. Secondly, 'the acquisition of information becomes an end in itself' (p 851). Pupils are seen as little more than tape recorders which store up information, 'isolated from action and purpose.' The pupil is not required to think through alternative approaches tc problems nor to consider the merits and disadvantages of applying different perspectives. They are taught skills independently of applications of real world situations. Finally the role of the teacher is that of a manager of learners rather than of learning. The teacher's job following this practice is to start and stop lessons, keep order, explain the rules and procedures and to ensure pupils follow them. In short the approach most commonly adopted by teachers is perceived to rob both the teacher and child of the opportunity to solve meaningful problems and to capitalize on enquiry modes of learning. Common practice, in this view, divorces participants from their natural intellectual strengths and from an appreciation of the proper structure of mathematical experience. In avoiding an examination of applications work it necessarily limits pupils' flexibility and leaves them to rely on rote solutions if they are fortunate enough to recognize a familiar problem-type. The only problems they can solve are the ones which have structures very similar to ones practised. Improving common teaching practice by developing the fine details of its component skills would, according to Rosenshine and Stevens, give significant gains on test scores. Unfortunately such gains are not the whole — or even a significant part — of what is wanted from a mathematics education.

Romberg and Carpenter's view of mathematics learning involves the development of enquiry methods in the face of real problems. Any step away from this challenge is a step away from mathematics learning. The obvious solution is not to refine common practice but to reorientate it by the use of teaching techniques which concentrate on real problems, germane to children's interests and demanding the development not only of procedures and concepts but also of strategies of enquiry and application. Such an approach does not claim to dispense with a need for children to learn concepts, facts and procedures. Rather, it makes the claim that these elements are best met and most easily remembered in the context of practical use on substantive problems.

Now this solution to the problem of mathematics teaching is as well known and has as long a history as the problem itself. Its major premises can be traced through James (1899), Dewey (1916), Brownell (1928), Polya (1957) to DES (Plowden Report) (1967) and Cockcroft (1982). In some of its varieties a purely enquiry mode of learning seems indicated (Dewey, 1916). In others, a more eclectic approach involving an appreciation of practice and a variety of teaching methods is adopted (Cockcroft, 1982). In whatever the precise form the crucial element is that in their mathematics education, children should be actively engaged in exploring mathematical technique in the face of substantive problems. Techniques should be discussed amongst teachers and pupils. Pupils should discuss ideas with their peers. Discussion should develop confidence in sharing and evaluating ideas. Problems should relate to real life issues meaningful to the pupils.

This approach to mathematics education has not been left at the level of rhetoric. It has been converted into practical guides (for example Nuffield, 1967) and presented to student teachers in initial training and to experienced teachers on in-service courses. Its virtues have been expounded frequently, confidently and enthusiastically. In the introduction to the Nuffield (1967) scheme, for example, it was claimed,

> Running through all the work is the central notion that the children must be set free to make their own discoveries and to think for themselves and so achieve understanding instead of learning mysterious drills. In this way the whole attitude to the subject can be changed and, 'ugh, no, I didn't like maths' will be heard no more. (p (i))

Despite the enthusiasm, industry and pedigree of its proponents this obvious solution to the problem of mathematics education appears to have had very little impact on the average classroom. In reflecting on its national survey of American practice the National Advisory Committee on Mathematical Education concluded, 'Teachers are essentially teaching the same way they were taught in schools. Almost none of the concepts, methods, or big ideas of modern mathematics programmes have appeared in the median classroom' (NACOME, 1975, quoted in Romberg and Carpenter, 1986, p 851). And only sixteen years after the fanfares announcing the British Nuffield Project it has been concluded that, '... mathematics educators are likely to look back on the "new maths" era as on a wild, and perhaps misspent youth, filled with energy and enthusiasm, idealism and naivete, but faintly embarrassing ...' (Hayden, 1983, p 1). No more than a very small minority of teachers had practised what had been preached (Page, 1983).

Why Don't Teachers Take Good Advice?

There is nothing even faintly wild, youthful or naive about the Cockcroft Report's recipe for good classroom practice in teaching mathematics. Regardless of the age of the pupils it is recommended that,

> Mathematics teaching should include opportunities for exposition by the teacher; discussion between teacher and pupils and between pupils themselves; appropriate practical work; consolidation and practice of fundamental skills and routines; problem solving, including the application of mathematics to everyday situations; investigational work. (Cockcroft, 1982, p 71)

In the very next paragraph it is observed,

> In setting out this list we are aware that we are not saying anything that has not already been said many times and over many years. ... Yet we are aware that ... there are still many [classrooms] in which the mathematics teaching does not include even a majority of these elements. (p 72)

Surveys of practice suggest that this is something of an understatement. Teacher exposition and pupil practice clearly predominate. Discussion, practical work, investigations and real life applications are at a premium. Why does this pattern persist in the face of good advice?

Cockcroft suggests that one of the reasons might be that brief statements on good practice do not sufficiently explain what is intended. The Report therefore develops each of the above headings and explores some of the practical applications. It is noted, for example, that investigational work might be set in formal projects familiar to primary schools but it might equally profitably have its origins in pupils' questions of the 'what would happen if . . .?' variety. Such investigations might be of very short duration but,

> The essential requirement is that pupils should be encouraged to think in this way and that . . . there should be a willingness on the part of the teacher to follow some false trails and not to say at the outset that the trail leads nowhere. Nor should an interesting line of thought be curtailed because 'there is no time' or because 'it is not in the syllabus' (p 74).

Cockcroft suggests that this and similar principles be the subject of extensive discussion amongst teachers in order to establish an appropriate rationale for mathematics teaching.

The suggestion that teachers have not been made sufficiently aware of essential pedagogic principles has also been put forward as an account of why the 'new maths' projects failed in the 1960s and 1970s. Hayden (1983) observed that the teachers' textbooks contained all the necessary subject content but largely omitted the pedagogical innovations that were the key characteristics of the programme. Stephens and Romberg (1985) have also argued that teachers lacked an appreciation of the philosophy behind the new mathematics and the values associated with its various activities. The net effect, suggested Kamii (1985) was that teachers became the, '. . . mere executors (if not the executioners) of someone else's decision.'

It seems to us unlikely that the failure of these innovations rests on the teachers' lack of appreciation of their underlying philosophy or rationale. At the heart of the call for more practical work, investigations and applications is a view of the learner as intrinsically motivated and inventive, a view of the syllabus as flexible and negotiable and a view of the teacher which leans towards power sharing rather than the autocratic. In general terms this model of learning, teaching and classroom relationships has been set before teachers for many decades. Since the popularization of Piaget's work the virtues of the model have been extensively advertised. It seems to us inconceivable that teachers are not aware of the claims made for it. The resistance to implementing the model may be seen to be broader and deeper than that.

In a broader context, researchers on teaching and learning are very familiar with the teaching profession's rejection, at the level of practical action, of their earnest advice. Glaser *et al* (1977), for example, noted that, 'At the present time, cognitive psychology's findings and techniques have not significantly influenced teaching practices, instructional processes nor the design of conditions for learning' (p 495). Ausubel and Robinson (1969) observed that despite favourable responses on the part of teachers to psychology courses, 'the behaviour of these same teachers observed later in the classroom, has typically shown distressingly little influence of the principle and theories which they had presumably learned.' (p iii). The lack of effect of educational precept on teaching practice is not confined to school teachers. Eisner (1983) showed that academics in his own University School of Education found it extremely difficult to give examples of how precepts based on research or analysis were practised in their teaching. Put bluntly, Eisner's survey suggests that even those who develop educational precepts do not, in any clearly articulated way, practise them.

Even sharpening teachers' awareness of precepts by the threat of assessment does not guarantee their implementation. In a report on a Mathematical Association Diploma in an eminent British university it was shown that only 29 per cent of the participating teachers introduced investigational work into their classroom even though it was a compulsory, assessable component of their course. It was not reported what fraction of the 29 per cent continued with investigations after the course was completed.

A hint that the problem runs more deeply than a mere lack of awareness can be taken from the observation that teachers frequently welcome and endorse the advice which they then fail to convert into practice (Duffy and Roehler, 1985). The issue has been put most poignantly by a teacher recalled by Brown (1968). The teacher mused, 'I don't know why I keep going to meetings to learn more about becoming a better teacher. I already know how to teach ten times better than I ever can' (p 9).

When teachers do not respond positively to invitations to introduce new or refined practices, it is rarely the case that they are unaware of the invite. Rather, it is more likely that they do what all humans do. They incorporate the information they receive into their own perceptions of reality. Teachers have been found to be, '. . . most receptive to proposals for change that fit with current classroom procedures and did not cause major disruptions' (Feiman-Nemser and Loden, 1986, p 516). In this light teaching practices prevail as part of the common culture of teaching because they are adaptive to the circumstances under which teachers work.

The conservative nature of teaching practices has been most persistently criticized for the general rejection of methods of teaching considered to foster the attainment of higher level learning goals (such as learning to learn, problem solving, learning strategies) and involving discussion or discovery methods of teaching which require the teacher to adopt a more reactive role. These processes have either not appeared or not survived in the average classroom even when the teachers have shared the curriculum developers' aspirations (Duffy and Roehler, 1985). The difficulties of establishing investigational or conversational methods of working in mathematics may be seen as instances of this general problem. Getting teachers to understand and share aspirations is only a very minor first step in establishing classroom practices appropriate to higher level goals. The second, and major step, is to understand the circumstances under which teachers must implement such practices. As has been recently stressed, 'Those who criticize teachers for maintaining this "practicality ethic" may underestimate the added complications that flow from attempts to alter established practice and the degree to which current practices are highly adaptive to classroom realities' (Feiman-Nemser and Loden, 1986, p 516).

Classroom Realities

There are those who long ago recognized that the teacher's job is not as straightforward as it might appear. William James (1899) observed that the science of pedagogy was,

> . . . much like the science of war . . . In war all you have to do is to work your enemy into a position from which the natural obstacles prevent him escaping . . . then to fall on him in numbers superior to his own, at a moment when you have led him to think you are far away; and so with a minimum exposure of your own troops, to hack his force to pieces and take the remainder prisoners. Just so in teaching, you simply work your pupil into such a state of interest in what you are going to teach him that every other object of attention is banished from his mind; then reveal it to him so impressively that he will remember the occasion to his dying day. . . . The

> principles being so plain there would be nothing but victories for the masters of science, either on the battlefield or in the schoolroom, if they did not both have to make their application to an incalculable quantity in the shape of the mind of their opponent. The mind of your enemy, the pupil is working away from you as eagerly as in the mind of the commander on the other side. (pp 9 and 10)

William James' image refers to a very simple model of teaching in which information is transferred from teacher to child. The difficulties he so vividly illustrates were raised to question the simplicity of the advice offered to teachers almost a century ago. James would doubtless have had fun with some modern precepts.

A century ago, teachers were not burdened with the problem of differences between pupils' levels of attainment or their differential accumulation of relevant knowledge. Nowadays this is taken to be a very serious issue for the teacher to attend to. So serious is this problem that one internationally famous psychologist has opined, 'If I had to reduce all of educational psychology to just one principle I would say this: the most important single factor influencing learning is what the learner already knows. Ascertain this and teach him accordingly' (Ausubel, 1968). This apparently simple dictum invites teachers to match assigned work to individual pupil's levels of attainment. To do this the teacher must necessarily diagnose the state of the pupil's knowledge. Now even under ideal circumstances this is excruciatingly difficult. After forty years of careful study under experimental conditions, psychologists still dispute the interpretations of children's performances on well known Piagetian tests (Brown and Desforges, 1979). The diagnostic programme facing a teacher with thirty children and looking for starting points in all aspects of her curriculum would be of such a scale as to ensure that she never got round to any teaching (Kuhn, 1979).

Estimates suggest that were a teacher's time to be shared out equally amongst all her pupils, she would have approximately ten minutes per day with each child in which to advance the whole curriculum. And, of course, the state of the child's knowledge is changing all the while so that diagnosis, even if satisfactory on one occasion, would need updating on another. Of course these observations of practicalities do not dismiss the sense of Ausubel's precept. But we can see at a glance that a teacher would have to make decisions which would, necessarily, drastically limit her diagnostic activity. Under the circumstances we would not expect all the work assigned to pupils to be well matched to their attainments. In fact our own research (Bennett *et al*, 1984) showed that in a sample of sixteen classes of 7-year-old pupils, 57 per cent of tasks set in mathematics were either too easy or too difficult for the children. The research illustrates just how difficult it is to practise one perfectly reasonable precept. Whether one is surprised or disappointed at the difference between aspiration and practice must depend on one's understanding of the force of classroom realities. Certainly any professional judgment of this matter must rest on an understanding of the situation (Desforges, 1985), as must any attempt to improve classroom performance. The above remarks do not constitute such an understanding. They serve only to illustrate the teacher's situation.

Much more comprehensive illustrations of classroom life from the teacher's perspective are available (Jackson, 1968; Smith and Geoffrey, 1968). Jackson's vivid account showed that teachers attend to a huge number of, often conflicting, demands. She must attend to individual children whilst monitoring the rest of the class, supply corrective feedback whilst developing confidence and give children time to think whilst keeping her eye on the clock. The teacher distributes time to activities and attention and physical resources to children. She organizes movement about the room, the composition of groups and the flow of events. Jackson estimated that elementary grade teachers take part in 200–300 exchanges every hour of their working day. Almost every exchange involves a unique mixture of personalities and circumstances. The teachers' role in these circumstances has been variously characterized as a 'quartermaster' or 'traffic cop' or 'ringmaster'. These analogies serve to convey images of a person imposing some order on a complex and potentially chaotic situation. The processes of establishing and maintaining order — in terms of a productive flow of activity and not to be confused with mere discipline or control — clearly occupy a great deal of the teacher's attention and time. In the eyes of many scholars engaged in classroom research these processes are the most salient aspect of the teacher's job and as such we might anticipate that they would have considerable influence on mathematics teaching. They have recently become the subject of extensive empirical work and theoretical analysis (Doyle, 1986).

Doyle has argued that all classrooms have certain features in common, 'There are . . . important elements already in place when teachers and students arrive at the classroom door' (p 394). All classrooms are inevitably crowded places serving many different purposes, needs and preferences; many events and tasks take place and these events are supported by a restricted supply of resources. This entails that difficult choices will have to be made, 'Waiting a few extra moments for a student to answer a question, can affect that student's motivation to learn as well as the pace of the lesson and the attention of the other students in the class' (*ibid*). A second significant feature of classrooms is that many things happen at once. 'During a discussion a teacher must listen to student answers, watch other students for signs of comprehension or confusion, formulate the next question and scan the class for possible misbehaviour' (*ibid*).

The pressures of multiple factors and simultaneous events are compounded by the rapid pace of classroom life. Added to this there is an unpredictability about classroom events. There are frequent interruptions and distractions and it is often difficult, '. . . to anticipate how an activity will go on a particular day with a particular group'. In the face of a rapid sequence of events, many of which are unforseeable, the teacher has very little time to reflect before reacting.

The classroom is very much like an arena. Events — especially those involving the teacher — are witnessed by a large number of pupils. If a teacher reprimands a child, a significant portion of the class may learn a sanction. By the same token, if she fails to reprimand a child, the class might learn something to their advantage about her management skills.

In this public domain the classroom participants witness, share and accumulate a common set of experiences which, in Doyle's terms, becomes the class 'history'. This influences the subsequent conduct of the class. In the compilation of the class history the very earliest sessions have been shown to exert an influence on events throughout the year (Emmer *et al*, 1980). Many experienced teachers place a particular emphasis on shaping up the class in their early meetings in recognition of their experience that to fail to do so will severely constrain their hopes for advancing learning.

Doyle suggests that at the heart of all these factors is a superabundance of information to be processed. Sources of information include books, displays, workcards, materials and a vast amount of verbal and non-verbal behaviour. The information-rich classroom environment is inhabited by pupils and teachers who share the restriction common to all humans of being able to attend to only a very limited amount of information at any one moment. In order to adapt to this environment they have to learn to make fruitful selections amongst available information and to make as many exchanges and procedures as routine as possible. Order has to be imposed on the information. The teacher's situation in this respect has been compared to that of the learner-driver (Desforges *et al*, 1985).

> The advantage of making actions automatic is that it reduces the load on information processing capacity or focal attention. This frees more attention for monitoring the environment. The process of automation may be compared with that of learning to drive a car. Initially, basic controls demand almost all the novice's attention. Once the basics are made automatic, attention is available for the road. When reading the road becomes routinised or automated, attention is available for conversation with the passengers. Perhaps this example illustrates not only the advantages but also the potentially serious disadvantages of making behaviour automatic. (*ibid*, pp 167–8)

Whatever the potential disadvantages, establishing routine and order are known to be enduring concerns for teachers (Clark and Peterson, 1986; Woods, 1976; Yinger, 1980). Teachers devote a lot of time teaching routines and classroom responsibilities (Cornbleth and Korth, 1983; Sieber, 1981). In the light of Doyle's account of the classroom the concern for order appears to be a necessary condition of classroom life if the teacher is to have any impact on the development of proceedings.

In Doyle's theory an important consequence flows from this conclusion. This is that to establish order, the teacher must have the cooperation of the pupils. With order and cooperation as first priorities, pupil's engagement with tasks must come second. This league of priority is frequently seen in class discussion lessons (Bennett *et al*, 1984) in which only a very few pupils take part whilst the rest act as, at best, an audience but more commonly as mere bystanders. In Doyle's terms, '... in the daily world of the classroom, order can, and often does, exist without full and continuous engagement by all students in learning tasks. Moreover, non-engagement is not necessarily problematic in establishing and sustaining order even though it may be unsatisfactory for learning.' (Doyle, 1986).

Doyle emphasizes that cooperation is a reciprocal process: it is achieved *with* pupils and it, '... depends on their willingness to follow along with the unfolding of the event' (*ibid*, p 396). Pupils are neither passive nor powerless in this matter. Cooperation is negotiated, albeit by subtle processes. Some tasks cause much more disruption — and hence threaten order — than do others. Several studies have shown that children offer a lot of resistance to intellectually demanding work (Atwood, 1983; Davis and McKnight, 1976). Jorgensen (1977) discovered that when the demands of assigned reading material fell below pupils' levels of ability their classroom behaviour improved. In Doyle's model, teacher and pupils are engaged in persistent reciprocal adjustments of work setting and delivery with cooperation, and order, potentially at stake. Thus the work set plays a crucial function in the negotiation of cooperation and hence order.

At the heart of this process of negotiation is an 'exchange of performance for grades'. That is to say the pupil tries to deliver what the teacher wants. This is not always straightforward because information in the classroom is often misleading or ambiguous. This was well illustrated in our recent study of top infant classes (Desforges *et al*, 1985) in which we observed a teacher giving a writing lesson to a class of 7-year-olds. Her intention was to stimulate their imaginations in order to procure from them a piece of creative writing. To this end she demonstrated a working model of a volcano which the children found extremely thrilling. It provoked a great deal of talk. The teacher then asked the class to, 'Write me an exciting story about what it would be like to live near a volcano.' In fact they did no such thing. What they actually produced was a very predictable — and generally very short — account of the demonstration. They took very great pains over writing these accounts in order to make their work clean, neat and tidy. Whilst they were doing so the teacher circulated round the class pointing out smudges, uneven writing and those children who had omitted to put the date on their work. She praised neat work. These were all very public messages. Exciting stories got no further mention. Clearly the children had set out to deliver what they knew — from the class history — that their teacher wanted.

In Doyle's terms they had used their 'interpretive competence' which is that sense or skill which enables pupils to select the information telling them how to get the teacher's praise. The teacher's praise or other rewards thus acts as proxy for what she evidently 'really wants' as opposed to her potentially, misleading instructions as to what she says she wants. In this respect, how the teacher assesses work becomes the pupils' clearest definition of what it is they have to do to cooperate with her. Doyle describes this situation as 'the assessment system driving the work system'. 'Accountability plays a key role in determining the value or significance of work in a classroom: products that are evaluated strictly by the teacher are more likely to be seen as serious work, that is, work that "counts"' (Doyle, 1986 p 406). In the above example, the teacher's persistent reference to neatness was taken by the children to be a clear indication of what she was serious about. In that sense her assessment behaviour played an important part in defining the task. More strictly, since her specific assessment behaviour came after the children had started the task, it was the history of her assessment behaviour which helped to define the task.

This view of the role of assessment helps to account for pupils' resistance to tasks with higher level cognitive demands. Such tasks, including demands for reasoning, understanding and problem identification for example, have no 'right answers' or predictable procedures. They present the pupil with high levels of 'risk and ambiguity' (Doyle, 1983) in the sense that it is not clear what to do to earn praise or reward. They present problems to the teacher too in the sense that she cannot necessarily see what to reward in order to sustain pupils' performances. If, for example, the teacher in the writing task reported above, had seriously concentrated on creativity and ignored production features (such as neatness) there would have been very little reward for anyone. It is difficult to identify creativity in process; it might involve staring out of the window for inspiration — a behaviour which is notoriously easy to confuse with disengagement and lack of cooperation.

In Doyle's analysis, tasks demanding higher level cognitive processes, and, particularly those that have no concrete, verifiable, 'correct' product, pose problems for the assessment system in the classroom. In doing so they challenge the 'rate of exchange' in which cooperation is negotiated and hence potentially threaten classroom order. In support of this analysis Doyle (1986) cites a study by Morine-Dershimer (1983) of recitation lessons in American early grade classes. Evidently, when the teachers asked convergent questions, (that is those with a specific, correct answer) the rate of participation was directly related to the ability of the pupils. When the teachers asked divergent questions requiring opinions almost all answers became acceptable and the task involved merely participating. High ability pupils became reluctant to respond and pupils paid less attention to each other's contributions. The findings in this study are consistent with the view that tasks high in ambiguity (that is divergent questions) devalue the familiar rates of exchange in classrooms and in so doing, threaten the cooperation of pupils. Pupils do not remain passive in this situation.

Several studies have revealed pupils' techniques for increasing the levels of predictability of tasks high in ambiguity. Mehan (1974) reported that 6-year-olds hesitated to give answers until the teacher had answered for them. Pupils are adept at luring the teacher into gradually increasing the amount of information until the answer is virtually provided. In a study by MacKay (1978) one pupil observed, 'Yeah, I hardly do nothing. All you gotta do is act dumb and Mr. Y. will tell you the right answer. You just gotta wait, you know, and he'll tell you' (cited in Doyle, 1983, p 184). This process of guiding the teacher has been called 'piloting' (Lundgren, 1977). It is doubtless instantly recognizable to experienced teachers but nonetheless difficult to resist.

In Davis and McKnight's (1976) study in which they attempted to shift the demands of high school algebra tasks from routines to understanding, the pupils' responses were much less subtle. They refused to cooperate and argued that they had a right to be told what to do. Davis and McKnight observed, 'it is no longer a mystery why so many teachers and so many textbooks present ninth grade algebra as a rote algorithmic subject. The pressure on you to do exactly that is formidable!' (cited in Doyle, 1983, p 185).

Studies in an English pre-school playgroup (Ward and Rowe, 1985) suggest that 4-year-olds, whilst less demanding, may be no less anxious to interpret and follow the assessment structure of the classroom. Ward and Rowe observed, for example, 'Several 4-year-olds working round a dough table. Each child was working independently of the others. And each worked on his or her unique project. One made a ring and offered it to the play leader. She said, "How lovely" and kept the ring on her finger. All the other children immediately abandoned their projects, made dough rings and presented them in the same fashion' (p 3).

The picture that Doyle develops from these and other studies is one of a subtle — and sometimes a not so subtle — process of task definition in which the accountability system is the vehicle of information and in which teachers and pupils play reciprocal roles. Doyle's theory is not an account of how pupils learn. Rather, it emphasizes that whilst learning is a covert — in the head — process, it takes place in school within the complex social world of the classroom and an explanation of how that world impinges on learning is offered. It is suggested that the well documented press of exchanges and events in the classroom is best conceived as a potentially massive overload of information. To adapt to such an environment it is held necessary to select information and impose order on the events. Order in this sense is not mere discipline but more generally an organized and predictable sequence of activities. Establishing and maintaining order becomes the teacher's essential priority. The tasks set in the classroom both influence and are influenced by the established order. Tasks with higher level cognitive demands increase the pupils' risks and the ambiguity involved in engagement and thus alter the commonly (and usually readily) established exchange rate in classrooms — that of an exchange of tangible rewards for tangible products. Pupils like to know where they stand. For this reason, tasks demanding higher order thought processes are resisted or subverted by pupils. Resistance puts cooperation at risk. Teachers are lured into or connive at subversion and higher level task demands are frequently re-negotiated in the direction of routine procedures. What are the educational implications of this analysis?

Doyle (1983) emphasizes that the accountability system in the classroom plays a central role in management of academic work. This poses a special problem for the tasks most valued by educationalists, that is, those demanding non-routine, higher level thought processes — which have been shown to be most difficult to manage and least evident in classrooms. In this perspective, curriculum, instruction and management are intimately interrelated.

If Doyle's analysis is correct, establishing and sustaining tasks demanding higher order thinking in the day-to-day curriculum will be a much more challenging undertaking than at first sight it appears. Doyle suggests that not enough is known about the, '... fundamental tension that exists between management and instruction ...' and that research is needed to get a richer picture of how classroom order and work are maintained. These studies, it is suggested, should incorporate information about academic work and social interactions, components which have classically been separated by educational researchers. In particular, suggests Doyle, efforts should be made to establish teachers' perceptions of management in order to provide a better foundation for designing teacher education programmes. In other words, significant classroom realities reside in the teacher's mind. Those who want to change classroom realities must first understand teachers' cognitive representations of them. It is essential to comprehend in what ways teachers' strategies of management and instruction are adaptive to their classrooms.

The Present Study

We see the problem of mathematics teaching as the problem of establishing and sustaining a curriculum demanding the development of higher order thinking, including the skills associated with problem identification, problem solving and the acquisition of the strategies needed to apply basic concepts and routines. It is at once the problem of identifying apposite skills, designing appropriate tasks and managing these tasks as part of a broader curriculum. It could be said that little is known about any of these aspects of the matter. Least of all is known about the implications for and effects of classroom management. What is certainly known is that it is a difficult problem as the resistant teaching practices of the last century indicate.

It has been frequently suggested that the reason why teachers do not change their practices in desired directions is that they are not sufficiently aware of the value and importance of higher cognitive objectives in education, of the philosophy underpinning them and of the notional procedures of how to attain them. As we have indicated, we have found this argument simply unconvincing. In our previous research (Desforges, 1978 and 1985; Bennett *et al*, 1984) we have found teachers very well informed on these matters. If hard work, a desire to do well by children and an awareness of goals were all that were needed, there would, in our view, be no problem of mathematics education.

A more plausible account of the limitations of traditional and enduring classroom practices is that they are the result of teachers' efforts to manage a situation which stretches their resources to the limits. This account is consistent both with our research and our teaching experience. Put in this blunt fashion current practice looks like an immovable obstacle on the path to a more challenging curriculum. The attraction of Doyle's account is that it is an attempt to understand the essence of the way in which teaching practice is adaptive to classroom circumstance and in consequence promises to provide a firm intellectual foundation on which attempts to alter practice might be based.

Unfortunately there is a number of limitations to Doyle's analysis. It is based on a relatively limited number of studies and whilst all the evidence quoted is consistent with the theory, there appears to have been no attempt to date to test it. Secondly, all the studies refer to the North American experience. Whilst we have shown that we share the American problem with mathematics education, it does not necessarily follow that our practices are similarly constrained. As we have already observed, social reality resides in the mind and is consequently the subject of cultural influences. When Doyle suggests, for example, that the, 'exchange of performance for grades' is a significant classroom process, we may suspect this to be a peculiarly American consideration — a resultant, perhaps, of the achievement ethic and the profit motive. Consequently, it is an open question as to whether Doyle's analysis of American classroom management practices is valid in Britain.

These reservations notwithstanding, Doyle's account is attractive. This is in part because it is the only account we have and in part because it is couched in terms of the limits of human information processing which limits know no national boundaries. It is sufficiently attractive to pose the following questions. Is it the case that the problems of classroom management produce and sustain the contemporary and limited classroom practices in teaching mathematics? If so, what do those problems look like to the teacher? What factors do teachers take into consideration in adopting management and teaching techniques and what factors force her to amend her goals or behaviours in the day-to-day world of the classroom?

Answers to these questions would help us to design more fruitful approaches to the enhancement of practice and it was to these sorts of questions that our research was devoted. In short, we set out with the assumption that teachers are aware, industrious people and we aspired to understand them as they taught mathematics.

A Note on Method

In order to meet our aims we decided that it was necessary to (a) describe teachers' actions in teaching mathematics to young children (this in turn necessitated describing their children's responses since we anticipated that much of the teachers' action is, in fact, reaction); (b) describe the teachers' thinking before, during and after mathematics sessions and from these descriptions (c) develop our understanding of how teachers manage the academic work of mathematics in the classroom through the identification of classroom realities as teachers experience them.

For our purposes it was essential to recruit a sample of teachers who were themselves interested in the business of mathematics teaching. It was crucial that the teachers were experienced and generally successful because we did not wish our classroom observations or subsequent discussion to be dominated by consideration of ill discipline or by the effects of poor organization. Our aims and research design demanded that we collect a lot of detailed data from each classroom and consequently it was necessary that we focussed on a few teachers in depth rather than lightly sampling the experiences of a larger number.

For these reasons we recruited seven experienced infant teachers. Our initial contacts with them were through their local reputations as teachers with a special interest and facility in infant teaching and with sufficient self confidence to be likely to survive being researched. Our subsequent work with these teachers bore out their reputations. As we shall show, we found them to be extremely industrious, devoted to their children and very aware of contemporary issues in teaching. Beyond that we make no special claim that these teachers are representative of any particular group.

The teachers worked in a variety of situations from middle class, rural to inner city decay. Between them they had ninety years' teaching experience. Their classes comprised two reception, one second year, two mixed second and third year and two third year infants. Five of the classes contained more than thirty children. The two reception classes contained twelve and twenty-two children.

Once contacted, the general aims and methods of the research were explained to the teachers. The seven were a sub-set of ten teachers contacted. Three rejected us, two on the grounds that they had particularly difficult children at that point and the third because she felt she could not cope with being recorded on video.

Each of the seven teachers was studied intensely for a period of three weeks. At the beginning of the period discussions were held with the teacher to familiarize the researcher with her classroom and mathematics programme. A description of the content she expected to cover with her class in the period of the research was also acquired. A small test was designed to cover this content and administered individually to a small sample of the children in her class to ascertain what they knew of the concepts and procedures before they met them with their teachers.

A period of observation then ensued in which a sample of mathematics sessions was recorded on video tape. The earliest recordings were seen merely as opportunities for the teacher and her class to become familiar with the equipment and the researcher. This was rapidly achieved. Several of the teachers informed us that they quickly forgot our presence.

As well as recording some sessions, field notes were taken of the activities of a sample of children. For these purposes the teacher chose six children spanning the range of attainments in the class. It should be noted that the precise levels and ranges of attainment of the children were not of great significance since our focus of interest was the teacher.

After the period of familiarization, each video recording was discussed in detail with the teacher and the discussions were tape recorded for subsequent analysis. In the discussion the teacher pointed out matters of interest to her on the tape and was asked to explain their significance and her thinking at these points. On a second run through the tape, the video was stopped at points sampling the beginning, middle and end of the session recorded. At each point the teacher was asked to describe and explain her thinking. If further points of interest to the researcher had not been captured in these two runs through the video, they were brought to the teacher's attention on a third, selective run. In this way 696 events were discussed. The tapes were discussed on the day they were recorded.

Clearly the majority of our data comprised teachers' memories of their thinking some two to five hours after the event, the memories being prompted by the video. The deficiencies of introspective data are well known (Titchener, 1912; Calderhead, 1985). Verbal accounts of thought processes may be more rationalization than rationality. Memories are known to be leaky, selective and creative. To that degree, reminiscences about thoughts which are a working day old may be considered to be less than perfectly valid. Whilst recognizing these reservations we can observe that our teachers' accounts were consistent with the events on tape and consistent within themselves. The teachers were frequently self-critical indicating no desire to create false impressions. The teachers were very sure of the accounts. We asked them, during discussions, to rate their confidence in their recall on a four point scale, (1) for certain to (4) for very unsure. Eighty per cent of the accounts were scored (1) and almost all the rest scored (2). Whatever the reservations we considered that if we wanted to know what the teachers thought we had better ask them.

Discussions with a particular teacher continued until she felt she had no more to say and was repeating herself. This occurred after the analysis of approximately one hour's video tape covering, typically, five teaching sessions in conversations lasting many hours.

The conversations provoked by specific events, whilst starting with an account of the teacher's thoughts at that time, ranged widely as the teachers placed the events in context as they saw it. In this way were obtained the teachers' views on teaching, learning, thinking, the curriculum, classroom processes, school constraints and particular children. Lest this spontaneous talk had omitted anything of significance, a follow-up interview was conducted with the teachers after the whole observation period of the research had been completed.

During the period in the classroom, diagnostic interviews were held with the selected children to ascertain their grasp of their mathematics sessions. Before leaving the classes, the children were given a brief, informal test mirroring the one given at the inception of the research. In all 296 interviews were held with the children.

In this brief account of our research method we have omitted the technical details. These can be found in Cockburn (1986).

Chapter 2

The Teachers' General Views

In our discussions with the teachers, most of the talk was about particular children in particular incidents. Frequently however, the discussions ranged more widely. In the conversations on both particularities and generalities, the teachers revealed a broad set of attitudes which, by their own accounts, permeated their thinking, revealed their priorities and informed their teaching at the level of both planning and action.

Amongst these schemas it seemed that, whilst the teachers recognized the issue of national anxiety about mathematics education, they were emphatically more occupied with the matters of children's learning, the broad range of relevant differences amongst their children, the choice and development of appropriate teaching methods and the constraints and pressures they experienced in their classrooms. Since, in the teachers' view, these factors influenced their minute by minute teaching it is important if we are to understand the teachers, that we have a grasp of such schemas.

In this chapter we sketch out the teachers' views on these general matters. In doing so, we have used the teachers' own words.

National Standards and Individual Children

All the teachers were well aware of the general and national concern about mathematics standards although they took very different views of the issue:

> It's absolutely justified. I think bad maths teaching starts in the first school. I've taught maths badly. I still don't teach it well. ... I think it's an absolute minefield.

> I can't think why someone has not stood up and said, 'standards have never been higher'. It's a convenient scapegoat — blaming education for the mess society's let itself in for.

> I don't know. Standards have always seemed high to me. The basics seem to have been put in. I've never come across anyone who has not been able to do the basics.

> I think the concern is justified. Secondary education is seriously at fault. First schools take a very practical approach to maths with a lot of success.

> I don't know whether it's justified. It's dangerous. I think it's caused a swing away from practical maths and towards these formal schemes and workbooks — especially with able children.

Whatever their views, the teachers were not terribly interested in this matter. Their attitude, as on all other issues raised, cannot be understood except in the light of the teachers' experience of the uniqueness of each child in their care and their recognition of the vast differences between children in their classes. These considerations permeated their thoughts about materials, teaching technique, their response to pressure and the organization and administration of their classrooms. Their conceptions of the differences between children were rich and diverse. They recognized dimensions of intellectual, social and emotional variety but whilst their descriptions of children were frequently illustrated by contrasts and comparisons, the teachers tried to stress that they were less interested in differences between children than in the unique characteristics of each child.

Intellectual differences between children are huge:

> I don't start the scheme until the top group is through the October half term. I want to get to know what they can do first. This year I have a boy who has finished the infant scheme and two books of the junior work; and I've got a girl who did not start the scheme until the middle of March.

> Some children never seem to need concrete experience. They grasp an idea immediately — others seem to need the props for ever. I doubt if some children ever do reach an abstract stage.

> Because they are so young it is easy to assume that they all need concrete experience. It's not true. Some seem to have real insight. Others work with concrete materials exclusively.

They need work breaking down to the smallest items and leading through it step by step.

At the beginning of the year I found one girl five books in front but she only understood half of what the others had grasped.

Attainment differences are compounded by social and attitudinal differences:

There's a huge spread at age 5. In attainment, knowledge, relationships. Some are very independent. Some are totally dependent.

They'll whizz through the cards with great confidence whilst others won't move without checking with you.

Some are lazy. Some are chatterboxes. Some chatter and still get the work done. Some toil and get nowhere.

There again you get some children who find getting a new concept very difficult but once they get it they'll work very hard with it. But then when they reach the next new concept they've slowed right down again. Then there are others who appear to be O.K. but when you peer below the surface they need a lot of propping up.

The individual relationship is crucial and you need thirty different styles to meet their needs.

Children as Learners and Teachers

Regardless of their children's different qualities, the teachers were fascinated — and sometimes mystified — by their ways of learning and their capacity to think about the use their experiences.

I don't know how they learn. They certainly don't learn by copying out of books.

I wish I knew how they learn. I haven't got a clue. When we get stuck on something you can almost see them thinking and I'd love to know what is going on in their heads.

You can't really say that a child has learned x, full stop. Children can forget — or lapse. They often seem to shelve things whilst they learn something else. They do it almost deliberately whilst they are busy taking something big in. It's almost as if they didn't know something at all — but it comes back later.

I don't know how they acquire the variety of approaches they use. They are very good at using real life experience in the classroom.

I don't know. But filling in books is one hell of a way not to learn.

They are very observant. It's wonderful when they come along with something they have seen — for example the symmetry in a clockface — it's terrifically exciting.

They come up with things they want to do all the time. The top class wanted to beat the *Guinness Book of Records*' daisy chain! They made one for two days and when they measured it they were miles out.

Children know a tremendous amount of things. In amongst a group of children you have a tremendous reserve of knowledge to draw on. I brought in a book about Roman soldiers. One lad suggested we make a fort. Another announced he'd seen a Roman fort on Hadrian's Wall. We had no shortage of knowledge to work on.

Children know so much that is useful to maths teaching. You have to try to keep in touch with their world. I try to watch children's TV and read their comics. It's not only important to the relationship — its vital when they start choosing topics for graphs!

Children do think about things, transfer ideas, do the abstraction and arrive at a conclusion — and they are well able to share their knowledge.

All the teachers were impressed by the capacity of children to explain and communicate their ideas to other children.

Children can communicate far better with each other than they can with you — or so it seems at times.

Sometimes a child has a good way of explaining something. If they are doing a workcard I can say, 'Have a word with Thomas. He's just done that.' It's good for Thomas, good for

> the child and it takes the pressure off me. It's not exploitation — it's a positive learning situation.

For this reason, these teachers were all keen on child-to-child talk but they recognized some severe difficulties in using this approach to learning.

> I like to get children to talk about ideas amongst themselves but it is a very difficult technique to manage. I've never been really good at getting it going. Children are able to share ideas but not in a set up way. They are much better spontaneously.

> Somehow you have to harness the interaction that goes on. There is interaction but I don't think — as a teacher — that you are always conscious of the fact that it is profitable.

> Sharing ideas amongst children is better than telling them. It gives them a much better feeling of competence. When children contribute they all seem to listen very hard — and they come up with a far greater range of ideas than ever I would think of. But using them as explainers is not so easy.

> Using children to explain or discuss ideas is good in principle — but whether it works in practice is a different matter. They can be very ego-centric. When we did a decision tree on taking milk and sugar with tea they were only interested in showing where their picture was. They were not interested in discussing the workings of the decision tree.

> If you are not there you don't know what's going on. Do they explain or do they just tell? It's a nice idea and some children do have an incredible facility to explain but I would not be inclined to put a bright child with a slower child. Slower children can be easily put off. They need an extremely careful dialogue with someone who can sense the underlying problem and amend the task accordingly.

> You really need an adult to sustain an activity. Children don't sense the progression.

> It's good for the talkers because verbalizing clarifies your thinking. It's difficult to tell how it helps the children who are not contributing.

> I don't think they necessarily listen. It's the talker who profits.

> You can't always get children to talk and pushing it always make it worse.

Whilst convinced of the quality of children's thinking, the extent of their knowledge and their impressive ability to state their view, the teachers had reservations about young children's capacity to adapt a message to the emergent needs of a learner, that is, about children's capacity to teach. All the teachers cast doubts on their own ability to manage child-child interaction as a classroom resource. Most went further and doubted young children's willingness to amend a message according to what they had heard — a fatal flaw since the teachers considered the capacity to listen and act accordingly to be the crucial teaching skill. They prized their own ability to do this:

> Trying to listen is the *only* way you can find out about a child's understanding.

> Knowing what to pick up comes from listening to children.

> You have to avoid leaping in. You have to listen.

> I never, never, never ignore a child who wants to say something. It's too important. Listening is at the heart of teaching and learning.

> Trying to find time to listen to all of them is the biggest challenge.

Listening is not only important for technical feedback. It secures the relationship considered essential for good teaching.

> What children have to say is very important to them.

> It all boils down to communicating. Would you want to relate to someone who did not listen? Could you?

In doubting young children's capacity to listen productively — or doubting their own capacity to manage child-child talk — they raised questions about the pedagogic utility of such exchanges.

General Teaching Methods

All the teachers claimed to have used a variety of methods for teaching mathematics and concluded that there was no 'best way'. Variety was essential if the needs of particular children were to be met. And variety needed a thinking teacher who could spot what was necessary:

> There is no best way. You are different and they are different. If you are still thinking you are bound to want to change ideas.

> There *is* a right method. It's called 'making the right judgments'. Is this approach appropriate for *this* concept with *this* child?

> Some teachers are still looking for *a* book like SPMG. And they're going to start at page one and go through to the end and that will teach their maths. And there isn't ever going to be anything like that.

> The mixture of techniques is imponderable. No two classes are ever the same. You play it by ear. You pick up the signs.

Of course, as already noted, listening is an important activity in 'picking up the signs'. And this, together with watching, was considered to be an integral part of all teaching activities. For example, several of the teachers were keen on exposition (although no teacher used that term: they referred rather to 'teaching') but only if its effects were closely monitored:

> Teaching is important — and it's missing in a lot of schools. I teach. Then I listen and watch them all to see that they have got it. If they haven't, I respond as needed.

> Teaching is dangerous. I don't favour it because children can sit there completely passively and they might not be taking in anything at all. It needs follow up and questioning.

> You have to do a lot of direct teaching with little ones. You have to establish the basics. Essentially it's talking and doing. But everybody talks and does. And you have to watch them.

> I always start with a demonstration and then ask them to do it practically. If it turns out they can't do it I spend more time with them — even if it's just one-to-one and the rest of the class have to stand on their heads for five minutes.

Practice was considered to be extremely important but only if it were matched to the needs of the children and its effects carefully monitored:

> Practice with practical materials is very important at this stage but I only have practice when I can keep an eye on them to stop them getting into bad habits.

> Drill and repetition are so necessary to some kids and can give them a real kick when successful.

> Practice is really important but some children need very little. But children who have a lot of difficulty getting a concept need a lot of practice. It's very daunting to be pushing on all the time. They need a period of success.

> Practice is important but you need to ring the changes a lot. A problem in real life is never going to be presented in quite the same way. You have to look at a given skill or concept from a lot of different angles. Practice can also get boring. It needs to be disguised repetition.

> You need as much time as necessary on practice. I don't want to sound mystical. Some children need a lot more than others. Some need the security of practice even if it's not stretching them. I'd be guided by the children.

Two factors were considered to be common to all the teachers' expositions and all their children's practice. These were first, the opportunity for teacher-pupil discussion and secondly the opportunity to do practical work.

> Practical work is absolutely fundamental for almost all children at this stage.

> Practical work is always the way in.

> All my teaching is really demonstrations using concrete materials.

And during the children's use of practical materials the teachers considered it essential to monitor activities and discuss problems.

> I like them to talk about what they are doing and why they are doing it.

> The talk is essential. It's not only feedback. It's fun. If we can't get it right we always have a giggle and say we'll try again another day.

The talk thus serves much more than a diagnostic function. Indeed, it would seem from the teachers' comments that if it served only that purpose it would fail in its other aspect — that of sustaining morale and facillitating a good relationship. In contrast to child-child talk, teacher-child talk gave the teachers confidence that they knew what was going on and that they were fulfilling their responsibilities to sustain progress in children's learning.

Teacher exposition, demonstration, practical work and practice, all closely followed up and monitored in teacher-pupil discussions, formed the bulk of the mathematics teaching according to the teachers.

Additionally, applications work was considered to be crucial.

> Of course it's important they learn to use what they know. And in a sense we are doing that all the time. At their level they are solving problems when they sort out the pencils, when we do the register together, when we look for patterns in shapes in the classroom and so on. They are all problems to them. They all involve applications of skills and knowledge.

> Problem solving involves using skills. We do a lot of word problems about real things — sweets and toys and so on.

> It's important that you set traps and problems — all in a fun sort of way. They enjoy the challenge. The adrenalin starts moving. They learn much more in that sort of way.

Whilst problem solving was considered essential and, in an informal way permeated the curriculum in the form of conversation about sharing materials, doing the dinner numbers and so, the teachers expressed reservations about the practicality of problem solving in the classroom and about some of the advice they had been given on this matter.

> We did some logical thinking with top infants this year. By all accounts that was fairly horrendous. We bought some work cards that looked very good and we gave them to children who had the basic skills and understanding. The slower children had dreadful problems.

> Making maths real through problems is not as easy as it looks. A real problem probably doesn't come from school. Thirty experiences cannot be brought into the classroom. If you select amongst their ideas, that's another problem.

> I went on a course where they talked about using real experiences and gave the example of the child who had helped his Dad build a chicken coop at the weekend and all the maths that could be got out of that. Now you cannot really expect the others to be interested in someone else's chicken coop.

> The real matter is to make the classwork as realistic as possible rather than trying to bring phoney reality — if you see what I mean — into the class.

> I do a bit. But there is usually more social interaction than maths. If they see some children doing graphs and ask to do a graph I let them of course. We agree something they are interested in — usually six choices people can make — and they organize the data and do the planning. They have to organize that side of it and I show them what to do to get a graph.

The relevance of problem solving to the curriculum was not contested. However, the teachers clearly had had difficulty in putting this on a footing that was in their view satisfactory.

Limiting Factors

All the teachers thoroughly enjoyed teaching young children. None of them thought that they worked under unreasonable constraints or limitations. That being said they made it plain that classroom life was far from ideal and that classrooms were not their idea of a superior learning environment.

> Kids really need one-to-one attention. If they've got problems with reading, talking, maths and emotional problems — as many of mine have — the whole job becomes very frustrating.

> The problem is too many children, too small a space and not enough time.

> In schools now it's quite expensive to make materials. There are limited amounts of card. You've got to be very selective.

> I would not make anything that's one-off. Everything has to be adaptable.

> There are some lovely ideas in resource books but the materials are so expensive.

> Whatever you do it's got to be something that is easy to manage.

> It needs organization. The teaching group has to be small whilst the rest are gainfully engaged. Children like routine so

> that they know where they are. On the other hand they like surprises and variety too. You have to get these into balance so that you have a routine into which you can feed variety.
>
> You have to have a system that creates time. It's a system that pins certain children to their desks whilst you concentrate on others. Although I don't think there's great value in filling in boxes, there is value in keeping my sanity.
>
> If you have small, busy groups, other groups create a distraction. Discussion needs a quiet time. I have tried to get groups to work and others to discuss. But it's distracting. My mind is not totally on what's happening because I've got to be aware of what others are doing. It ruins everyone's train of thought.

Clearly the teachers are vividly aware of the limits of their most precious resource of time and they are equally aware that their various solutions to this problem have their own disadvantages.

Despite the teachers very positive attitude to children and their appreciation of children's thinking capacities, they were conscious that children were not always their allies in developing mathematics skills. Each class, the teachers reported, contained children who were ever willing to cause disturbances and, unless they were very well organized, one teacher reported, 'about 50 per cent of children would do as little as possible'. A more pervasive problem recognized by the teachers, was that many children were very anxious to please and, as a consequence were preoccupied with identifying the procedure which made the teacher happy rather than thinking about what they were supposed to be doing.

> They just try to make it come right for me. They love to please you. They really do.
>
> They try to reach the answer that they thought I wanted. All children do it. It's difficult as a teacher not to make them think you are after a right answer because very often you are anyway.
>
> Something that is always there [is] 'what does *she* want?' and that's precisely the thing you are trying for them *not* to do. But they must from their point of view. You set the task to the group so it's reasonable that they assume you have a purpose. You *are* dictating what's going on in a sense. So the major thing is 'what does *she* want' and it's very difficult to avoid.

The teachers did not spontaneously suggest that they were under any other kind of pressure which might detract from the quality of their mathematics teaching. When it was suggested to them that heads or parents or their colleagues might make demands for higher standards (which in turn might entail rushed teaching) they considered such involvement to be entirely legitimate and no cause for concern.

> I've had to give explanations of what we are doing to parents. But that's reasonable — that's not a threat.

> Teachers should be able to justify what they are doing. If they cannot justify it they should not be doing it.

> The head takes a close interest in what we are doing and we do NFER tests. But I presume everything is OK.

> The head monitors progress all the time. He knows what's going on. That's his job.

> Parents are often anxious for their children to do well. You have to explain what you are doing and why. You have to win their confidence.

Whilst recognizing the legitimate interests of many parties, all the teachers had had unfortunate experiences in which particular people had, in their view, behaved improperly in the extent of their demands and in so doing had caused upset for the teacher and problems for the child.

> You have to do what you think is right. Some parents are extremely opinionated. They push them on to hundreds, tens and units. Buy workbooks. They don't realize how much confusion they cause. I resist that sort of influence completely.

> The middle school head wanted our children to be on such and such a workbook. We take the children as far as they can go but it is a real dilemma.

> We had a great to-do with a mother (who was a teacher) who was convinced her child was underachieving. She wanted him on hundreds, tens and units and he couldn't make a set of eight objects. I battled with her all year but I wouldn't give in.

> In my last school we had three parallel classes. One of the teachers pushed everyone through the scheme. She didn't do the practical work — which I did. I was getting further and

further behind. That was real pressure. I didn't want to let my kids down. I felt dreadful. But I did the practical work.

In summary the teachers recognized that there are limitations, constraints and some pressures on the quality of the experiences they can provide for their children. In particular they felt the shortage of time, space and equipment and occasionally they felt the pressure of people with ill-considered interests in children's attainment. They did not see any of these forces as insuperable and had long ago learned to manage them. With respect to resource limitations they recognized their management was less than ideal: it was, however, practical. With respect to the demands of external accountability, they considered this entirely reasonable so long as those asking the questions were prepared to accept the teachers' professional judgments.

Although aware of the constraints and difficulties under which they worked, the teachers were equally conscious of the children's strengths, their intellectual power, and their capacity to learn, think and attack problems. They had a great deal of respect for children as learners — a respect not based on psychological theory but on years of experience. The teachers also recognized, albeit ruefully, that children's thinking powers were frequently used in counterproductive ways — specifically, to make what Donaldson (1978) has called 'human sense' of the social situation by attempting to read the teacher's mind in order to please her by giving her what she appeared to want. Such applications of sense were not typically what the teachers said they wanted.

Whatever the teachers' appreciation of these processes, they were sensitive most of all to the unique nature of individual children and indicated that this individuality was manifest in the child's social, emotional and intellectual response to classroom events. Two equally advanced children, for example, might pose completely different challenges to the teacher. As we shall see, these differences were never far from the teachers' minds.

To meet the challenges posed by their children, the teachers articulated a need to deploy a range of teaching strategies — including most of those named in the Cockcroft recipe of good practice — each to be closely monitored in its effects.

In establishing the high level of awareness of the processes of teaching and learning amongst the teachers whose practice we studied we are not seeking to generalize these comments to the rest of the profession. We seek only to establish an important foundation from which to understand the practices we observed.

Chapter 3

Content and Management

In this chapter we move from the consideration of the teachers' general views on teaching and learning to the examination of the mathematics content they taught and their strategies for managing mathematics in their classrooms. Since we shift towards more specific facets of their teaching, it becomes necessary to take more cognizance of the teachers' particular circumstances. Also, from now on, it will be useful to identify each teacher with the situation in which they worked. Table 1 shows some pertinent details about the teachers and their classes. To preserve the teachers' anonymity we have used a code.

We trust these details are useful for subsequent reference as we describe and analyze the teachers' work. We have to recognize, however, that they do not, and we cannot in the these pages, convey the differences in these teachers' working conditions. Some of the teachers, notably Mrs. A. and Mrs. B., worked in very pleasant, well equipped surroundings showing signs aplenty of parental support and cooperation evident both at the classroom door each morning and in financial help for school funds. In stark contrast, Mrs. C. worked in an area of high unemployment and broken families. The area evinced a generally tough aura.

What the teachers had in common was rather a lot of children in a room no bigger than three times the size of an average lounge. Mrs. B., who had an uncommonly small class, had a correspondingly tiny room. She also worked as head of the school as well as the reception class teacher. What the LEA giveth the LEA taketh away.

Table 1: The teachers and their classes

Teacher	*Status*	*Approx age*	*Type of school (size)*	*Catchment area*	*Year group taught*	*Pupils in class*	*Average age*	*Predominant maths scheme used (subsidiary)*
Mrs. A.	Scale II	40	First (medium)	Rural	Mixed second and third years	33	6; 10	SPMG (Peak)
Mrs. B.	Head	50	First (small)	Rural	Reception	12	4; 4	Nuffield (Cormack and Fraser)
Mrs. C.	Scale II	40	First (medium)	Council estate	Reception	22	5; 6	Nuffield
Mrs. D.	Deputy head	40	First (medium)	Inner city	Third year	35	6; 11	SPMG
Mrs. E.	Deputy head	40	First (medium)	Council estate	Second years	35	5; 11	SPMG
Mrs. F.	Scale II	30	Primary (large)	Suburban	Mixed second and third years	34	6; 7	Peak (SMP)
Mrs. G.	Scale II	40	First (large)	Council estate	Third year	31	6; 8	SPMG

Mathematics Content

The children in the study were observed working on a wide range of concepts in the familiar British infant school mathematics curriculum. They were, collectively, seen to cover the following topics: counting, ordinal and cardinal properities of number, addition, subtraction, weight, length, time, capacity, money, shape, relational language and the concept of place value.

They met these concepts in discussion, practical work and in the form of workcards and workbooks. These latter materials dominated the children's mathematics' time. The reception class children were seen to be working on cards (or equivalent pencil and paper exercise) in 70 per cent of our observations, whilst for the 7-year-olds this ratio was 90 per cent.

Work on the cards was frequently supported by artefacts such as blocks, counters, clock faces, water-play materials and other pertinent objects permitting a concrete approach. The concepts the cards treated were almost entirely drawn from commercial mathematics schemes. It seemed clear that these schemes had a significant impact on the content of mathematics education in these classes. In this respect, these teachers conformed closely to the practices we had observed in our previous research in infant schools (Bennett *et al*, 1984).

That being said, most of the teachers in the present project were not uncritical of their commercial schemes. Indeed, some of the teachers had serious reservations about them. Mrs. A. made the following points on different occasions,

> I would never have bought this scheme in a million years ... would you have a scheme like this? ... It's not that it is absolutely bad, it's uninteresting. And it's not bad only because children are quite resilient and they don't mind what we do because they trust us ... Schemes make such heavy weather of things. They seem to batten hatches down in case something goes adrift.

Mrs. D. observed that, 'Schemes save the teachers from thinking. You need a system for keeping teachers thinking ... People who design maths schemes look at them only from the point of view of maths. They are insensitive to the differences between children.' Mrs. C. had a similar view, 'Schemes are dangerous. There is a real danager that they absolve you from decisions about appropriate work and sequence.'

Mrs. B. thought that there was a danger in almost any structured approach, 'Children over interpret the social context. They associate concepts with and only with the way *we* present them. Schemes are not clearly based on how young children learn', a point of view with which Mrs. D. clearly concurred, 'A scheme should start with how you think children learn maths.'

Less cogent, but nonetheless very general, criticisms were voiced about the schemes. Mrs. C., for example, felt that, 'There is too much in the scheme and it's too hard, too abstract. The language completely stymies them.' On the other hand, Mrs. F. found the scheme insufficiently thorough for her suburban children, 'There's too many holes. We do all the scheme and a lot more besides.'

Despite some reservations the impact of the commercial schemes was very evident. The major argument in their favour, at least in the eyes of these teachers, appeared to be that they provided some continuity of experience through the school. As Mrs. A., the schemes' most savage critic, indicated, once continuity in this form was opted for, you were rather stuck with the details, 'It's school policy. If they don't do the workbooks, when they go into another class they will be well down in the workbook scheme of affairs. The way the school works it demands this. It would be awful to be a really well informed child and go into the next class and be put on a book for a low achiever.'

Whatever their views, and it should be noted that Mrs. E. and Mrs. G. made no general or fundamental criticism of the scheme, all the teachers claimed to alter and/or supplement their main scheme to meet the needs of their particular classes.

Mrs. A., for example, suggested that whilst her children followed the SPMG workbooks to the letter to meet the school's policy, all their real mathematics was done opportunistically in art, craft or wherever it appeared on the curriculum. Mrs. F. and Mrs. G. decided at the start of the year what concepts they wanted their children to cover. They then used their central scheme and other available materials as a quarry from which to build a suitable programme of work. In Mrs. F.'s words, 'if I think there is something they ought to do I add it. Just because it's not in the scheme does not mean to say I shouldn't teach it. I get ideas from Fletcher, Scottish Primary. I use the Nuffield teachers' book. I write out what I want the children to cover and then look at the schemes. I don't think you should teach the schemes. You'd never cover the holes.'

The other teachers stuck predominantly to one scheme but used it very selectively. Mrs. C., for instance, said, 'I use the teacher's handbook a lot. I refer to it for ideas. I don't follow the order of the scheme. I pick and choose according to what I am doing. You couldn't follow it anyway. The language gives them the most horrendous difficulties.'

The teachers were not cavalier in their selections from the scheme or in their alternations to activities. Their amendments were based on their experience with children. For example, Mrs. C. decided to teach 'shape' in an order different from that advocated by the scheme. She explained, 'They say in Nuffield that you should introduce the solid shapes first because they feel that children — as a result of their early childhood experiences — come into contact with three-dimensional things more frequently and commonly than they do with flat shapes. But the children seem to find it much easier to understand flat shape names than solid shape names. I have found in other years that if I introduce solid shapes first, they find the vocabulary complex and unfamiliar. It's true they live in a 3D world but they are used to describing it with 2D language. They know 'round', 'circle', and 'square' but the solid shape names are completely alien to them.' Her decision to alter the order of approach was taken with considerable anxiety in the face of the expert knowledge held to be embodied in the scheme.

All the teachers made innumerable references to specific deficiencies of elements of the published schemes. These deficiencies were manifest to the teachers in the form of confusing workcards or impractical ideas. There is a tome to be written about teachers' detailed critiques of commercial schemes. Our present purpose, which is to give an image of the teachers' impressions of their working materials, will be served by a few illustrations.

Mrs. A., for example found that,

> . . . the visual aids in the books quite terrify them because they make it look like something difficult when in fact it's easy. With place value, for example, they are faced with diagrams of bundles of pencils and sticks, lots and lots of pages with an enormous number of lines. And children look at it and they say, 'I can't do that!' And I say, 'Yes you can because you don't need all that because you know it already'.

Mrs. F. thought that,

> . . . some of the language is ridiculous. This card [Peak, stage 2 showing photographs of buttons and bobbins] asks them to distinguish between the number of buttons and bobbins. The kids are totally unfamiliar with the word 'bobbins'. They

> cannot read it because the reading demand is far in advance of their attainments. You explain it to them but by the time they are back to their place the word 'bobbin' and the word 'button' are remarkably similar to them. A little bit of maths is buried under an awful lot of complication. It's almost as if it were designed to confuse!

All the teachers pointed to cards in which the reading demands were way in advance of the mathematics concept.

Equally, all had reservations about recording. Mrs. B., for example said, 'Recording pre-occupies the children. The amount of attention that goes into colouring, joining up points and so on is about a hundred times the amount of attention demanded by the maths on the cards!' And Mrs. D. noted, 'They do learn something from recording. I am not saying it is mindless. But it's not maths. It would be revolutionary to say the child was not going to write anything down.'

Many cards were seen to be impractical and sometimes foolishly so. Mrs. F. again, observed that, 'The capacity cards in his scheme leave a lot to be desired. There is one that says, How many spoonfuls of water will fill a bucket. The quantities involved are completely beyond their experience to estimate and they can't seriously imagine you would want to check it empirically.' Mrs. D., faced with introducing the link between addition and subtraction had, following her scheme, aspired to having children have a bag of marbles to experiment with. Somewhat fatalistically she observed, 'This is a good idea in theory because you can really see the processes at work. But you can't keep boys away from marbles. Before now I've put marbles into stiff polythene bags and sealed them permanently. Children chewed the corners and got them out. Marbles create chaos.'

The teachers reported a large number of problems with the apparatus associated with schemes. Many of their younger children had evinced difficulties in fixing Unifix cubes together, especially in quantities of more than three. Much of the children's time was thus spent in the struggle to manipulate materials rather than in the struggle to think about concepts and relationships. Some of the mathematics games for younger children appeared, in the teachers' view to rely on a social maturity that their children did not have — the ability to take turns or to lose gracefully for example. The teachers did not see these as insuperable problems because, typically, the children altered the rules to suit themselves but it appeared that the mathematics concept at the heart of the game was, in this way lost. Other games demanded manipulative skills that many of their children had yet to acquire. Several games, it was said, required children to cover sections of a board with counters or markers. Once the board had been nudged or brushed with an enthusiastic sleeve, the general state of the game and much of its mathematical point had been lost.

In the face of these various difficulties the teachers removed or re-designed cards — and sometimes whole sections of their schemes.

Despite their reservations, even the most critical teacher made extensive use of commercial schemes. Whatever thinking guided the teachers' selections and amendments, most of the children's assessed mathematics work appeared to come from such schemes. Mrs. E. followed her scheme closely. Mrs. B., C. and D. followed their schemes selectively, Mrs. F. and G. extended their core schemes using other schemes to make a programme. Mrs. A. followed her scheme to the letter but supplemented it with opportunistic mathematics teaching.

Whilst more than willing to criticize the schemes, the teachers gave no account of their attractions (even though these must have been more than sufficient to outweigh the deficiencies) save for Mrs. A.'s mention of the purpose the scheme served in providing a framework for establishing continuity of experience through a school. We are left to assume what the other attractions might be. One advantage might be that, designed by experts in mathematics education, the schemes confer status on the definition of the contents of a mathematics curriculum.

The teachers did not see themselves as mathematics experts and it might be that they took confidence from the scheme that they were providing their children with experiences which were of some significance as judged by mathematics educators. The teacher might take further support from the knowledge that each commercial scheme is followed by many thousands of schools. Both the status and generality of the schemes might serve as a form of curriculum justification to auditors — especially parents.

Another attraction might be that the schemes present a rich variety of ideas for treating concepts and processes; a variety that any particular teacher would be unlikely to invent. This attraction is implicit in the teachers' reference to the schemes as 'quarries of ideas'.

Perhaps the most understandable attraction might also have been voiced implicitly. All the teachers referred to the need to keep many children productively busy — preferably working at their own level of attainment — whilst the teacher worked with a particular group of children. Such a goal necessitates a prodigious amount of preparation of activities to a degree that would be almost impossible were a teacher left to her own resources. It is noteworthy that commercial schemes come complete with large quantities of material support. The materials are attractively designed and produced to standards beyond the reach of the class teacher. The materials help to solve the management problems created by the attempt to provide a diversified curriculum.

Managing the Mathematics Schemes

Operating a mathematics scheme presents the teacher with a number of management problems. Some of these stem from the necessity to arrange supplementary materials and to organize the children's access to the workcards. Other problems arise from the huge range of individual children's levels of attainment, their different levels of independence and their social behaviour.

Performance on schemes entails the use of pencils, crayons, paper, rulers, rubbers, counters, cubes, sand, water, scales, containers and numerous other materials. Some of these materials are in short supply and must be shared. Some of the activities are potentially messy and noisy and the whole business must proceed in a restricted space equipped for all other aspects of the curriculum. The smooth operation of mathematics work requires a great deal of organization and forethought to ensure that materials are present and accessible and that children exercise independence and mutual respect in their use. The seven teachers observed were exemplary in these respects. So smooth was the running of their classrooms that it was difficult to imagine that there might be a problem. Perhaps one way of appreciating the teachers' achievement in this respect is to consider the problems of getting just one 5-year-old to organize himself in his own home. One thing is certain and that is that any deficiencies in the availability or accessibility of materials and any uncertainty as to their relevance or use is bound to have a damaging effect on the quality of the children's mathematical experience. We were not able to study these effects because, as already noted, the teachers evinced no such problems.

As well as managing materials the teacher has to respond to the problem of her children's different levels of attainment and rates of working. The performance differences in mathematics in a class of similar chronological age are known to be enormous. Such performance differences were observed in this project. In Mrs. B.'s class Andrew (4; 3) could, on entry to school, count a string of thirteen objects, identify numerals up to ten and do simple calculations in addition if the objects were illustrated. In the same intake Darren (4; 1) could not count beyond two and had no other discernible mathematical attainment. Darren's low level of attainment was matched by his slow response to instruction. This is shown in the following exchange.

(Four children are sitting round the table with their teacher. She hands them each a card comprising the following:

1 2 3 4 5)
• • • • •

Teacher: Now then I want you to try writing the number underneath like that one there. (Mrs. B. points to the '1' on Darren's sheet.) Right, do you know what to do?
(Rebecca, Paul and Craig nod and quickly get down to work. Darren spends several seconds hunting for his pencil and then watches the other children for two minutes.)
Teacher: Come on Darren. You try and do that number one.
(Darren poises his pencil over '2')
(Thirty seconds later, Darren is still poised as his teacher returns to him.)
Teacher: No, start here (Teacher points to dot beneath '1') and then go down.
Paul: I've done it Mrs. B.!
Teacher: Clever boy. Right go and do your writing. How are you doing Craig? Well done, that's a nice three.
(Darren has just gone over the '1' on the worksheet.)
Teacher: No, not over it, below. Do it where I've done the dot. Look here Darren, you start there. (Teacher points to dot beneath '1'.)
Rebecca: Done it Mrs. B.! Done it Mrs. B.!
Teacher: Well done. Now go and do your writing.
(Darren completes copying '1'.)
Darren: I've done it.
Teacher: Have you done it? Right can you do a number two?
Craig: I've finished Mrs. B.!

The slow rate of progress can be enduring. Mark entered the same class as Thomas and Darren. Initially he had no sense of numerical quantity. After two and a half terms in school he still had trouble as shown in the following extract. (Mark and Mrs. B. were working with a box of toys containing cars, boats and trucks of various colours and sizes.)

Teacher: Take out three blue cars. (Mark takes out six cars of various colours.)
Teacher: Take out four red boats. (Mark takes out five boats of various colours.)
Teacher: Take out three yellow trucks. (Mark gets out seven objects.)
Teacher: How many did I ask you for?
Mark: Five.
(Teacher puts out three green cars.)
Teacher: Get me the same number of blue cars. (Mark gets out five blue cars.)
Teacher: How many have you got?
Mark: One, two, three, four, five, six (poking six times in the general direction of the objects.)
(Lest distracting interpretations of Mark's performance should enter the reader's mind it should be emphasized that he was not physically impaired in any way.)

Children at similar levels of attainment work at very different speeds. The four 7-year-olds who were set the card shown in figure 1 were of very similar attainment. A preliminary interview showed that each knew the answers and how to proceed. Jennifer completed the task in six minutes whilst Elizabeth took twenty-five minute. These differences were partly accounted for by speeds of writing, partly by application to the task and partly by levels of confidence. Jennifer wrote quickly, stuck to the task without distraction and was sure she was right. Elizabeth wrote slowly, was easily distracted and distracted others with social chit chat and took pains to check that she had the same answers as everybody else.

The differences illustrated in the above examples were typical amongst the children observed and present the teacher with the enduring problems of deciding on the appropriate level and pace of work for each child, and of monitoring their academic progress and engagement with the task set.

Amongst the seven teachers studied there were three different predominant strategies adopted to these problems. All the teachers reported that they had tried a variety of strategies and recognized that any option was associated with different advantages and disadvantages.

Mrs. A. opted mainly to individualize mathematics instruction. Her children worked at their own rate through the scheme and reported to her if in difficulty or for marking and further teaching on

Figure 1: A work card given to 7-year-olds

SMP
7-13
Length
7-2
Lizzy is tall and thin
Bob is tall and fat
Roy is short and thin
Peter is short and fat

Copy and fill in the gaps.

1. The tall children are called ____________ and ____________.
2. The short children are called ____________ and ____________.
3. ____________ is fat and tall.
4. Lizzy is ____________ and ____________.
5. You cannot be tall and ____________.
6. Peter is shorter than ____________ and ____________.

the completion of a section of work. Mathematics was done by the whole class daily. In these periods there was a steady stream of children who were either in difficulty or ready to progress. These children reported to Mrs. A.'s desk for attention. She operated a system of two queues — the fast queue and the slow queue. Children reported to the fast queue if they were having difficulty or if Mrs. A. required them to so that she could persistently monitor their work. Children reported to the slow queue on completion of a section for marking and further teaching. All the children understood and operated the queuing regulations. The teacher had told them that, 'the queue is a moving set of brains and we can all help each other.' In practice, in attending to a particular child the teacher involved other children in the immediate vicinity. The teacher recognized that standing in a queue was a potential waste of time. In view of this she arranged exhibitions and activities on the top surfaces of the storage drawers along which the queues stood. The children were encouraged to examine, use and discuss the materials and objects on display. These materials comprised, for example, mirrors, magnets, construction toys, balances with problems set and microprocessors (for example, *The Little Professor*). The queues stood in such a position that Mrs. A. could see the rest of the class. The children worked on their mathematics throughout the designated period. Fast workers did more mathematics, slower workers did less. All the children were allocated the same daily mathematics period. Their application was persistently monitored in this period. All their engagement with the teacher was at their own level for their own needs. At the level of strategy, all the children were allocated the same time and each was allocated the appropriate attention. In these senses the approach was just. It begs the question of whether the lower attainers need a lot more time allocated.

In complete contrast to Mrs. A., Mrs. C. and Mrs. G. predominately took a whole class approach to their teaching. Both teachers were very negative about queues. They considered queuing time as wasted time. Rather than teaching the same point over and over again to different children these teachers introduced work to the whole class but they had different follow up strategies. After Mrs. C.'s 5-year-olds had taken part in a class discussion they dissembled into groups. The children were not grouped in terms of ability but in consideration of friendship and work compatibility. Only one of these groups would be working on mathematics whilst the rest took up a broad range of curriculum activities including, for example, art, play and writing. The mathematics group would take up work related to the discussion. Unlike Mrs. A., who sat at her desk, Mrs. C. moved amongst all her children. There were no standing queues. Children remained at their work places and put up their hands when they required attention. Typically, at any one time, there would be half a dozen children signalling for attention. This is tantamount to a queue with the added problem that the teacher has to decide who has priority. Children who had completed recorded work had to get Mrs. C.'s attention for checking. If the work was acceptable they were sent off to a supplementary activity (for example the book corner or the play house) which did not require the teacher's detailed attention. Group activities were rotated in sessions throughout the day so that most children did some mathematics every day.

In this strategy the teacher has to present her work only once. Obviously it raises the question of what the children of different attainments make of this presentation. It seems likely that it will be easy for some and too hard for others but the teacher can respond to this problem as it emerges in practice by quickly circulating round the work groups. This involves more explanation but a lot less than one explanation per child.

Time was not the critical feature in the way Mrs. C. allocated mathematics. It will be recalled that Mrs. A. allocated equal time in which different children did very different amounts of mathematics. Mrs. C. allocated equal work (that is, all children got the same activity) but they took very different times over it. This system restricted the diversity of attainment in the class. Presumably the cost was that it held up more able children.

Mrs. G. had specific mathematics sessions. Like Mrs. C. she started with a discussion or demonstration involving the whole class. Unlike Mrs. C., all her follow up activities were mathematical. She organized her children in three attainment groups and usually designed for each group a separate activity judged appropriate to their attainment. The activities were all related to the content of the class session. For example in one session observed, the class of 7-year-olds had discussed quantities in the region eleven to twenty. They had identified number names, labelled the quantities in sets of objects and practised ordering and making numbers one more and one less than a given number. At the end of the discussion the teacher said:

> Now there are two different lots of work today. There is different work in each book but it is along the same lines.

[The teacher showed a page on which she had written

11 12 13 14 15 16 17 18 19 20

16 ——— and 1 more is ———→ □

11 ——————————————→ □

14 ——————————————→ □

12 ——— and 1 less than ———→ □

17 ——————————————→ □

15 ——————————————→ □

and so on.]

If you see there is a number line in your book it is not one to ten but eleven to twenty. The work in the book is about more than and less than. Michelle, what does it say?

Michelle: Sixteen and one more.
Teacher: Sixteen and one more. Well we did that on the carpet. What is sixteen and one more?
Kerry: Seventeen.
Teacher: Seventeen, so that's the number you put in the box.

Right folks, if you haven't got work in your books I've decided you can do them on your own. I'm going to put some on the board and you can write your own out.

[The teacher wrote calculations of the type

14 ——— and 3 more is ———→ □

11 ——— and 4 more is ———→ □

16 ——— and 2 less is ———→ □

on the board.]

Mrs. G. asked those with workbooks to get on and took the slowest attainers aside to do practical work on the same task as that allocated to the other children.

In one sense all the children were working on the same concept. By altering the task Mrs. G. had adjusted the recording aspect of the task demands. Some children had to take the material off the board, others had to write out only the answers while one group avoided the problems of recording by working orally with the teacher. Mrs. G. observed that, 'It took me ages to put all the sums in the children's books but it was essential otherwise the business of writing them out would have really messed the children up.'.

This system was operated daily. There were no queues. Children who wanted help or their work marking raised their hands. Those who finished to Mrs. G.'s satisfaction moved on to supplementary activities which typically took the form of reading or art work. This system requires considerable pre-planning both to prepare the different mathematics activities for the different levels of children's response and to prepare the supplementary activities — especially in art. As with Mrs. C., who also operated a whole class system, the children are allocated equal work rather than equal time.

The third and most common strategy for organizing mathematics teaching was adopted in the main by Mrs. B., D. and F. They organized their classes into attainment groups. They taught only one group at once. The groups followed the same scheme but at different rates and hence on a particular day they would be working on different contents. For Mrs. D. and F., with their classes of older infants, mathematics sessions were designated in which the whole class was involved on their different group topics. Thus, as with Mrs. A., all children were allocated the same quantity of mathematics time and what distinguished between the groups was their rate of working. Very large differences were evident between the groups both in the quantity of work done in a particular session and in the levels of attainment in the schemes. In Mrs. B.'s class of 4-year-olds the groups did not all work on mathematics at the same time and those who were doing mathematics had to complete the assigned activity rather than-work for a specific amount of time. On completion they moved on to a non-mathematical activity. There were no queues in any of these three classes. The teachers circulated to monitor and mark the children's work and this was done mainly in response to those with their hands up.

The group approach avoids the major disadvange of individualization (which is associated with having to explain everything to every individual child). It also avoids the major problem of class teaching (which arises from matching one explanation to the vast range of attainment levels in the class). Equally it fails to capitalize on the good points of these strategies .

As intimated earlier, the above strategies were those predominately operated. In fact each teacher occasionally used the range of techniques. Mrs. A., for example, sometimes conducted a demonstration for the whole class and sometimes (albeit rarely) organized the work for a group of children. For the above six teachers however such divergencies from the main strategy were infrequent. In contrast, Mrs. E. operated a distinct mixture of whole class and individualized teaching. For her class of middle infants she split the curriculum into number and the rest of the scheme (for example length, time, weight, relational language), she taught number to individuals in the same fashion as Mrs. A. and taught everything else on a whole class basis in a method similar to Mrs. C.

The teachers' critical problem in designing these strategies was how best to respond to the widely different mathematics attainment amongst their children. This is partly a question of allocating their time and partly a question of judging the level and pace of work assigned. In describing the above strategies we have noted how time was allocated. It was emphasized that all the children in each class were working on the same programme and it might have been inferred that the low attainers got the same work as the high attainers only at a later time in the school year. This was emphatically not the case. We have already seen that in the case of Mrs. G. all children met the same concept at the same time but that the children's response to it was tailored to different groups. In the other classes the low attainers did meet the same workcards as the high attainers later in the year but their route was frequently made simpler by the use of supplementary materials, additional explanations and more support on the task from the teacher.

We have already noted the large amount of logistical work which must be conducted effectively if mathematics work is to proceed smoothly. It is worth making the additional point that work on all the rest of the curriculum is fostered in the same room and, frequently, at the same time. The successful mathematics teacher needs to be a ruthlessly efficient quartermaster. The job would be virtually impossible if the children were not enabled as soon as possible to be independent operators with a social conscience.

For smooth work the children have to know where materials are, how to use them, how and when to share materials or take turns and how to tidy up. Even the youngest children in this study were effective in this way. In subsequent analysis this will receive no further comment because the teachers' success made it unremarkable. The teachers however could never take management for granted. It was never a problem. But without constant pre-planning and attention it would have soon been a major dislocating factor.

Chapter 4

Classroom Events: The Outsiders' Perspective

The aim of our study was to try to understand the teachers as they taught mathematics. We aspired to comprehend why the teachers did what they did and to grasp what effect their actions had on their pupils. Such an understanding must arise from a description of what apparently happened in the classrooms. Accounts from children, teachers and the researchers might all contribute to the image of classroom events: each view would be necessary for a full picture. As we noted in chapter 1, we attempted to obtain at least aspects of these several perspectives. Ideally, in describing classroom life, these perspectives should be treated simultaneously in order to retain the integrity of the original events and the vitality and interrelatedness of participants' interpretations. Such a technique requires the skills of a Tolstoy. Lacking this skill we have opted to present the researchers', the teachers' and the children's perspectives separately and to attempt a subsequent, integrated interpretation.

In this chapter we describe classroom mathematics from our own perspective. In doing so we must remind the reader of the old research adage, 'a perspective is a way of seeing: by the same token it is a way of not seeing.' As we shall see, what was at the periphery of our vision was at the teachers' focal point.

In developing our view we had the luxury of time in which to examine, and thanks to the video, to re-examine, events. We also had the freedom from having responsibility for these events as they occurred. At our disposal we had copious field notes on almost six months spent in the seven classrooms. We had hours of video tape and records of nearly 300 interviews with children to reflect on. In the space available here we can present only a brief selection from this data base and to do so we have had to impose some patterns on our observations. We asked ourselves, 'what were the salient events, as we saw them, in these classrooms?'

We perceived striking similarities across the classrooms. Whilst the teachers had very different personalities and taught children of markedly different attitudes and attainments, it seemed to us that mathematics sessions fell into two major categories evident in all the classrooms observed. In the first type of activity, children were either working on pencil and paper exercises from schemes or were being carefully prepared to work on such exercises. In 70 per cent of events observed in the reception classes children were so occupied. The ratio of these events rose to 90 per cent for the older children. In the second category (comprising 15 per cent of all sessions observed) the children were engaged in discussions with their teacher on topics which were not, in any close way, associated with pencil and paper exercises. In contrast to these two dominant patterns of activity, we saw one instance of mathematical activity proceeding under the navigation of children. This instance was salient in its rarity and its quality. In the following sections we discuss these three forms of mathematics activity namely, paper work, discussion and children's investigation.

Paper Work

The great majority of the children's mathematics time was spent being prepared by their teachers for, and subsequently carrying out, exercises from workcards or in workbooks. What was striking about the work was the business-like manner in which the teachers operated and the pleasure and industry evinced by the children.

The teachers' behaviour appeared to conform closely to the model of direct instruction described in chapter 1. It had all the advantages of directing the children to salient features of the work and of spelling out the desired ways of operating. In terms of Rosenshine's model of explicit teaching, the teachers' behaviour appeared to be exemplary. Unfortunately, direct instruction is associated with certain disadvantages (previously described in chapter 1) and these were also evident in the classrooms observed.

Almost invariably the children settled quickly to any work assigned. They worked with great industry and attention. Comparisons of the teachers' instructions and descriptions of the children's on-task behaviour showed that in 73 per cent of cases the children did exactly as they had been instructed. For example 7-year-old Neil had been taught that in the addition of tens and units he was to make up the numbers with rods, add the units and then add the tens. He appeared to follow this routine in his work. In a post-task interview his own account of how to do these sums was as follows. (The specific problem discussed was thirteen add twenty-five.)

> I have got to get thirteen out and then I have got to get twenty-five out and then I add the units and then I add the tens. I can do them in my head.

The interview showed that Neil could indeed do in his head all the sums he had been set. However, he invariably adhered to the procedure he had been taught.

Interviews conducted in this study confirmed what many researchers have found; young children frequently have strategies for successful calculations which are less time consuming and less Gothic than those taught by their teachers. Most children, however, conformed to taught procedures. Conformity to the teacher's routines was certainly not achieved by old fashioned fear. The procedures were not drummed into children using dread and pain. They were taught by more subtle processes which influenced the information handled by the pupils.

These processes appeared to permeate the mathematics sessions observed. The sessions typically followed a set pattern. There was an introduction to the work, key points or procedures were identified, there was a period of teacher-pupil interaction in which key points were exercised, a summary by the teacher of the procedure, a written exercise for the children and finally an assessment of the children's work. As the following paragraphs will illustrate, conformity arises out of the teacher's management of these sequences.

An introduction to a task might serve many purposes. It might, for example, pose the children with a problem, it might be used to call up and revise relevant knowledge and in this way link old work to new work or it might serve as a direct entry to a new task. In all the introductions observed there was a very rapid focus on the precise procedure to be dealt with. Sometimes this was immediate and direct. For example Mrs. F. introducing a new card said,

> This sheet is about months of the year. Now along the top there are the months written out for you all right? Starting with January which is the first month in the year going all the way down to December. Right now in the bit underneath you have to put in the right month.

Sometimes introductions were more discursive. In introducing some measuring work for example Mrs. D. proceeded as follows:

Mrs. D.: Now can you remember right at the beginning of this year we did some measuring. What did we measure?
Dean: Shoes.
Mrs. D.: Shoes, can you remember what we did?
Dean: We took off our shoes and put them on the scales.
Mrs. D.: We weighed them, what else did we do with them?
Mandy: We used Unifix.
Mrs. D.: How did we measure them?
Mandy: By getting them level (demonstrating the action of scales with her hands).
Dean: In tens.
Caroline: By Unifix.
Mrs. D.: That was when we weighed them. We measured them as well.
Mandy: On the tables.
Mrs. D.: We measured them on the tables.
Mandy: Some were small and some were short.
Mrs. D.: That's right, some were longer and some were shorter. We found out whose was the longest foot in the group.
Mandy: Mine.
Mrs. D.: Well this morning we're going to measure another part of our body. We are going to measure our span.
(The lesson continued with the teacher describing how to measure a span.)

Even at their most discursive, the teachers identified the specific procedure involved within two minutes. Whatever the children's interests and detours, their attention was directly focused on a specific procedure in this time. Teachers' apparent single-mindedness in their drive to the specific issue is well illustrated in the above example. Despite her open-ended first question Mrs. D. was not really interested in the children's immediate memory of weighing shoes. Rather she passed over this interest in the drive to get to her point. For all the use that was made of the children's contributions they might just as well not have been asked. It might be inferred — although it is not made explicit — that the teacher had a definite goal and was intent on focussing the children's minds on that, and only that.

Without exception introductions were followed up with an elaboration by the teacher of the key point of the activity. This involved describing the routine to be followed and going through it at least twice and more commonly four times. For example Mrs. B. instructed her class of reception children

> So on this paper you've got to find the ones that have two things in. That's the first row. All the things that have two things in. Okay? . . . Now remember on the first row you find all the ones that have two things in. Right?

And Mrs. C. (holding up a feely bag) told her class

> Now in this bag today I haven't just got solid shapes. There are flat shapes as well. You feel the shape and decide what's in it. It might by a square or a round or a cylinder. You feel the shape and see if you can think what shape it is.

With older children the key points became a little more complicated but the principle of repetition remained the same. Mrs. F. was working on coin equivalencies with her class of 7-year-olds. She said to them:

> I'm going to give you ten pence and you're going to change my ten pence using the coins. Using 5p's, 2p's and 1p's. Right, so, if you've got a ten pence, you've got to think of a way of changing that ten pence for the other coins. You are only allowed to use 2p's, 5p's and 1p's.

Mrs. F. gave some examples of the procedure emphasizing each time the starting point with the ten pence and the use limited to the five pence, two pence and one pence. She then asked several children to give an example and reminded the children in each instance of the procedure.

In the following instance Mrs. C. was teaching addition to a group of her reception children. They were sitting round a demonstration card which was marked out as shown in figure 2.

> *Mrs. C.*: Tony can you put three shells in there? (pointing to first circle)
> (Tony does so.)
> *Mrs. C.*: So what number have we got to put down here? (points to first square)
> *Tony*: Three.
> *Mrs. C.*: Now can you add six shells here (pointing to second circle)

Figure 2: Mrs. C.'s demonstration card

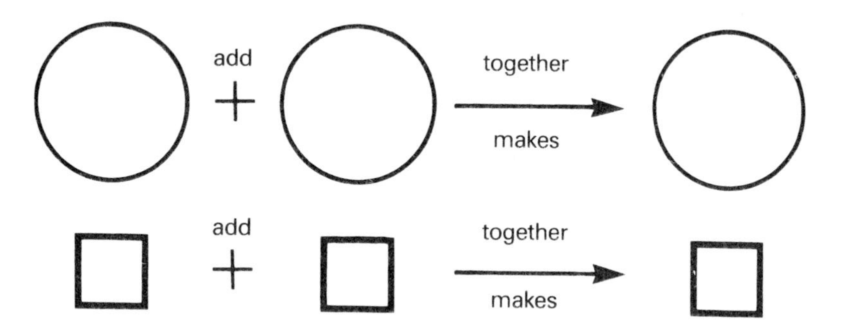

(Tony does so.)

Mrs. C.: So what number goes in here? (pointing to second square)

Simon: Six.

Mrs. C.: Now you push them altogether into here (points to third circle) and we count how many we've got. (The teacher pushes all the shells to the third circle and the children count as she picks them out.)

(Teacher clears the sheet.)

Mrs. C.: Right Michelle will you pick out five animals to go in here ...

And the whole routine was repeated several times.

It should be noted that preliminary interviews had shown that two or three of these children could do this kind of sum in their heads and some of them could do them using their fingers by counting on. The teacher in this session was showing a new routine. Other procedures were not considered. The teacher was apparently interested less in establishing the answer to a formal addition problem than in establishing a new routine.

The straight and narrow path through the workcards was followed in teacher-pupil interactions. A common feature of all the sessions involving oral work was a period of question and answer exchanges usually lasting several minutes. The teacher asked all the questions and the pupils supplied the answers. The questions referred only to the procedures to be followed and often had a special emphasis on recording the answers.

The children participated in these sessions with considerable enthusiasm. A constant sea of hands waving in appeal for an opportunity to respond was evident. The pupils seemed to be as committed as the teacher to following the routine. Very little spontaneity was witnessed in their reactions. In only 3 per cent of all the children's contributions to these sessions was there any unusual comment. When novel contributions were made they were briefly entertained before the teacher returned to the train of question and answers. For example, Mrs. B. was holding up fingers to be counted and told her class that most people had ten fingers.

Joanne: Mrs. B., I've got twelve fingers.
Mrs. B.: Are you sure? Have you counted them?
Joanne: Yes.
Andrew: I've got eighty-one.
Mrs. B.: Eighty-one. Mmm. Eighty-one is a lot.

As Mrs. E. used questions to take children through the routine of weighing:

Robert: I've been weighed in a box.
Mrs. E.: You've been weighed in a box? What do you mean — a box?
Robert: At the doctor's. A big box.
Mrs. E.: Yes . . . well . . .
(She then proceeds with her lesson.)

Such spontaneity as there was, was not punished. Rather, in the pursuit of procedures it was made marginal. That there was so little of it is perhaps best explained not by consideration of how it was received but by the realization of how it was cut out at source. One of the teachers' apparent preoccupations was to inhibit the children's shouting out their comments since this would clearly prevent her from locating her questions. Thus, all responses were expected to be direct answers to direct questions. What, in this regime, might be expected to be unusual was not the teachers' reactions to spontaneous remarks but that they should occur at all. Spontaneity was thus the privilege of the most inspired or the least socialized.

In the further preparation for paper work there appeared to be two major types of teacher-pupil exchange: one that focussed on what to do and the other on how to do it.

Instances focussing on what to do seemed to be a direct, oral pre-run of what was to be written. For example in getting her children to complete a workcard on long and short Mrs. F. proceeded as follows:

Mrs. F.: Right now then card 7–3. 'Some of these people are tall and some are short'. You have to copy the sentences and complete them. Who are the three tall people? Jennifer?

Jennifer: A., C. and F.

Mrs. F.: That's right. Who are the three short people? Eleanor?

Eleanor: B., D. and E.

Mrs. F.: That's right. Now look at the houses. 'Some of these houses are wide and some are narrow.' Again, copy and complete the sentences. Which are the three wide houses? Elizabeth?

Elizabeth: A., C. and F.

Mrs. F.: That's right and which are the three narrow houses? Sasha?

Sasha: B., D. and E.

Mrs. F.: Well done.

Such sessions typically went very smoothly. There were few errors. The children seemed familiar with the material to be practised.

Sessions focussing on how to do routines whilst following the same pattern of teacher questions and pupil responses often ran less smoothly. For example Mrs. D. introduced her class to a new routine for 'take away' problems:

Mrs. D.: Right now we are going to do take aways. Do you remember (writing on the blackboard as she says) four take away two leaves two ($4 - 2 = 2$). Right, now we are going to look at some sets.

(Draws (8/8 / 0/0) on blackboard.) Who can tell me a number story for that?

(Several hands go up.)

Barnaby: Two subtract . . .

Mrs. D.: Oh! (looks shocked) who can tell me a number story for this?

Katy: Two subtract two leaves four.

Mrs. D.: Two subtract two leaves four (writes $2 - 2 =$ ___ on the blackboard.) Let's look at it shall we?

Stewart: Nothing.

Mrs. D.: It should be nothing so it isn't that number story is it? What do you think Caroline?

Caroline: Two take away . . . Two take away one leaves four. Two take away one leaves one.

Mrs. D.: Two take away one leaves one (Teacher looks at the blackboard.)
Darren: Two add two.
Mrs. D.: What do you think Mandy?
Mandy: Four take away four.

At this point the teacher appeared to decide that the partition line in the set was confusing the children. She rubbed it out of her original diagram and took a new tack . . .

Mrs. D.: What have I got ? A set of what?
Mandy: Four.
Mrs. D.: A set of four. Now we are going to pretend it's a bag of sweets and your Mummy says that you can take two sweets out of it, how many will be left in the bag?
(Lots of hands go up.)
Barnaby: Two.
Mrs. D.: Two. So four take away two leaves two. Who can tell me how to write that number story down?
Lee: Two take away two.
Mrs. D.: Two take away two. Barnaby?
Barnaby: Four.
Mrs. D.: Four because that's how many there were in the bag to start with. Four take away two leaves two. (Mrs. D. writes this on the blackboard as she says it.) How many sweets did we start with?
Several children: Four.
Mrs. D.: Four. So before you can take any away that's the number you've got to think about. Let's look at another one.

The teacher then went step by step through several examples of starting with the set and partitioning it to indicate subtraction. She showed the children how two subtraction sentences could be written from a picture of a partitioned set and continued.

Mrs. D.: Good, well done. Shall we just try one more before I give you some to do.
(Mrs. D. puts (OOO OOO OOO / O) on the blackboard.) Right you have a turn Darren.
Darren: Ten.
Mrs. D.: Good, ten. (Mrs. D. writes '10' on the blackboard.)
Darren: Take away one leaves nine.

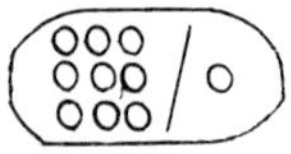

Mrs. D.: Good boy. Who can tell me the other one? Katy, how about you?
Katy: Ten take away nine leaves one.
Mrs. D.: Fantastic.

The teacher had achieved what she set out to do and had checked, with some of the children at least, that they had grasped how to handle the new layout. Pre-task interviews had shown that the children could do familiar subtraction in their head with the numbers in this range. Their initial errors in this session presumably arose from not knowing what the teacher wanted. The teacher recognized that there was some confusion and altered her somewhat exploratory approach in order to establish appropriate responses to a new format.

The initial errors (which it will be recalled included one child saying 'Two subtract two leaves four' and another child saying 'Two take away one leaves four') might be taken to indicate that the children had a very shaky grasp of subtraction. If this were inferred it would seem unwise to burden the children with esoteric forms of representation and it raises the question of what the children made of this experience. In interviews conducted after the above session it was clear that the children had not grasped the new routines. Even those who had performed well when the teacher conducted them could not interpret the partitioned set in the way the teacher had intended. It can be concluded that the children were somewhat confused by this session even though they responded well to the teacher's questions.

The teacher's own behaviour was consistent with the view that she set out to establish a routine and that problems on the way did not alter her goal but simply modified her approach. That there was initial confusion was obvious to her. It can be inferred that she saw this as a little, local difficulty rather than anything more profound.

The above example well illustrates the family of events in which teachers meet initial confusions which might give an outside observer cause for concern about the children's grasp of mathematics. The teachers however treat these as minor obstacles on their way to a predetermined goal. It has to be said that the cases in which the children faultlessly followed the routine in these interactions outnumbered the confusions in the ratio of about four to one. The significant point for this discussion however is that in the confusing events the children's errors and difficulties were rarely explored or discussed. They were set aside in favour of establishing a routine. This tactic can be seen as another process in establishing procedural conformity.

After the introductions, demonstrations and discussions the children were most typically set a written task. In the great majority of cases these tasks were very similiar if not identical to the tasks which had been used in discussion. When the children started their written tasks the teachers monitored their work either by circulating amongst them or by requiring the pupils to come to their desk. In most of the observed cases the children's work produced correct answers. In the cases where the teachers spotted errors the teachers redemonstrated the required approach and repeated these demonstrations until the child could perform the task. For example, 6-year-old Gemma was given an exercise of missing number sums. She had to fill in empty boxes in number statements. Her work was as follows:

$$1 + \boxed{4} = 5$$
$$6 + \boxed{2} = 8$$
$$5 + \boxed{4} = 9$$
$$\boxed{2} + 3 = 5$$
$$\boxed{6} + 1 = 7$$
$$\boxed{9} + 4 = 5$$
$$\boxed{7} + 3 = 4$$
$$6 + \boxed{10} = 7$$

At this point Mrs. F. interrupted Gemma and the following exchange took place.

Mrs. F.: Well these ones are okay (Mrs. F. ticked first five sums) but you've done these the wrong way round. Let's look at this sum (pointing to $\square + 4 = 5$). Now remember you are not adding this number (pointing to '4') and this number (pointing to '5'). It's what you must add to *this* number (that is '4') to get to that number (that is '5'). (Mrs. F. got five cubes.) You want five but you've only got four. How many more do you need? (Mrs. F. put a stick of four cubes by a stick of five cubes.) You want five but you've got four, how many more do you need?

Gemma: One.

In this fashion Mrs. F. went through five more examples of this type until she was convinced that Gemma had mastered the necessary procedure.

It was noted earlier that despite having alternative processes in the children's intellectual repertoire there was a huge degree of conformity to the teachers' strategies as the children did their written work. This was so even when — as was mainly the case — the teachers' processes were more elaborate than the children's private skills.

Much of the above discussion is an account of how such conformity comes about. Teachers open the sessions by identifying the process to be used. They explain and demonstrate the routine to be followed and use question and answer techniques to repeat and check on the uptake of their explanations. With the teacher so firmly in charge there are only two ways in which the exploration of alternative approaches might be initiated. One is through the spontaneous interruption by children offering their own perspectives and the second is by the examination of children's errors. The first alternative is managed away. In order to orchestrate the sessions the teacher requests that children do not shout out. Good manners therefore requires that the children speak only when spoken to and they are spoken to only to be asked a closed question. The second alternative is set aside. Children's errors are taken as signs of confusion and normally, rather than exploring this confusion teachers finds ways round it to present their intended routine. The teachers' processes of instilling a routine may thus be seen to be pervasive. They start by focussing attention on the task in hand and finish in a kind of mopping-up operation when checking children's work after the exercises.

As we noted, the teachers' sequence of behaviours conforms in most respects to Rosenshine's model of idealized direct instruction. This, perhaps, should come as no surprise. As we indicated in chapter 2, the teachers held this form of teaching in high regard and their many years of experience had doubtless contributed to make them expert at it. The teachers had described three key phases in this approach — input, participation and checking for take up (although they did not use quite these terms) and the record shows that their actions were as good as their words. Indeed so skillful and persistent were they at this form of teaching that the question of interest became not 'why do children conform to teachers' routines?', but, 'why do they not do so?' For in 27 per cent of observed cases the children did not use taught strategies in their written work.

Alternative Strategies in Paper Work

In half the cases in which the children did not follow the teachers' routines, the reason seemed clear: the children could not do the work because they lacked the prerequisite skills. These cases are examined in chapter 5.

In the remaining 13.5 per cent of cases, post-task interviews showed that the children could follow the teachers' routines but they did not do so. Occasionally this was due to errors arising from the misuse of immediate past experience. For example, 4-year-old Michael had become very familiar with the task of joining up dots to form number symbols. He was presented with a work sheet containing the following task:

1 2 3 4 5
• • • • •

and his teacher had required him to copy the numbers starting at the dot in each case. Michael joined up the dots.

In another case the children had spent the previous several days learning the names of three-dimensional shapes having previously dealt with two dimensions. As a revision the teacher set out a sheet of paper before them and asked them what shape it was.

Michael: Cuboid.
Mrs. C.: (outlining paper with her finger): What shape is that?
Antoinette: Cube.
Mrs. C.: It isn't a cube.
Paul: Cuboid.

The teacher redescribed the names of two- and three-dimensional shapes.

Commonly children's alternative strategies arose less from the intrusion of immediate past experience and more from what might be called smart thinking. For example Mrs. F. took Eleanor's group of 7-year-olds methodically through a batch of five workcards. Several minutes later Eleanor was seen to be rapidly recording her answers without any reference to the Unifix cubes she normally used. When asked how she was doing these sums Eleanor said, 'Well Mrs. F. has just told us them (that is the answers), hasn't she?'

In Mrs. A.'s class Robert took each new problem and looked to earlier pages in his workbook for where he had done it before. He then copied the answer from this source. This strategy actually took Robert a lot longer than doing the problems by the routine with which he was adept. However it has the enormous advantage of definitely giving the right answers.

A more common, smart technique was to find the secret of the page. It is commonly the case in infant mathematics schemes that a workcard exercises a particular number link. An example is shown below:

Add

6 + 3 = □	7 + 2 = □	2 + 7 = □
4 + 5 = □	9 + 0 = □	7 + 2 = □
8 + 1 = □	3 + 6 = □	4 + 5 = □
5 + 4 = □	8 + 1 = □	8 + 1 = □

It was not uncommon to find children treating these as a kind of code cracking exercise. If, as in the above example, the first two answers are nine they simply wrote '9' in all the other boxes. This strategy has a certain appeal and yet whilst not uncommon it was far more frequently the case that children worked laboriously through each element of such a card.

In a few instances children appeared to exhibit creativity in their mathematics work. For example Gavin was given the following card:

Figure 3: Gavin's workcard

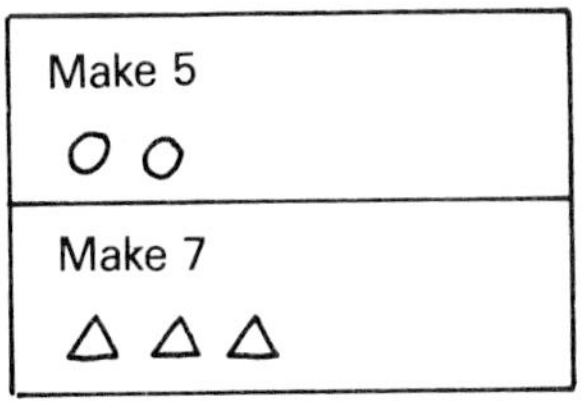

Make 5
O O
Make 7
△ △ △

In response to the instruction 'make 5' he drew a big circle round the three triangles and the two small circles.

Seven-year-old Kerry was given the task of adding tens and units with carrying. She had been shown the technique of adding the units, jotting down the carrying figure and then adding the tens including any carried. She ignored this in favour of her own strategy which involved, working in her head, adding the tens, adding the units, adjusting the tens as necessary and writing down the final answer. With this method she achieved 100 per cent success. When her teacher noted the lack of carrying figures she commented, 'Good, but could you just put the ones down there because one day you might forget them'. Again, the teacher's strategy is to encourage conformity.

The flame of independence however flickers on as shown in the case of Gemma. She had taken part with a group of other 7-year-olds in a demonstration of various ways of making up ten pence. She was then set problems exemplified in figure 4.

Figure 4: An example of Gemma's task

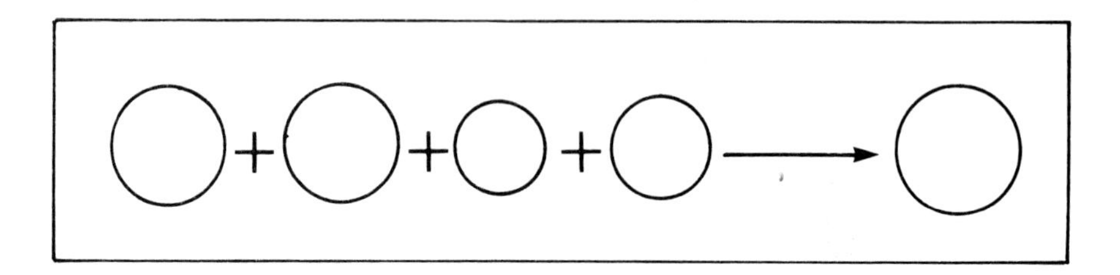

The teacher explained, 'I have actually drawn round the coins to help you. You've got to tell me how much each coin is worth.' The other children in the group set about fitting coins into the teacher's circles to obtain the required answers. In contrast Gemma ignored the circles, calculated ways of making up ten pence and inserted her solutions. In the example shown for instance she wrote '5p' in the first two circles and left the other blank. The teacher praised the others for their work but said to Gemma, 'You've got the answers but you've got the coins in the wrong places.' Gemma whispered in the field-worker's ear, 'some people think I can't do these . . . but I can.' This reaction perhaps gives us our only clue as to how even modest levels of inventiveness survive. It would appear that only sheer cussedness is proof against the pressures of conformity explicit in the techniques of direct instruction.

Discussions

It will be recalled that the younger children were engaged with work-cards in 70 per cent of the observed instances and the older children were so engaged for 90 per cent of observed instances. The majority of workcard sessions had involved a preliminary discussion on how to do the work. In six of the thirty-five sessions video recorded, work-cards did not make an appearance. These sessions were entirely oral and were orchestrated by the teacher. We have followed the teachers' term for these events and called them 'discussions'.

When doing their workcards the children typically sat at a table with three or four others. Whilst there was some occasional chattering amongst children at a particular table the pupils were predominantly occupied with their own work. For oral sessions the teacher normally gathered a group round her on the floor. According to the teachers, the purpose of this arrangement was to enable children to share ideas and learn from each other. Whilst the seating arrangement adopted might be described as intimate, most of the children spent most of their time paying attention to the work in hand. That being said there was a certain amount of distracted behaviour especially with the younger children. This included whispering, fiddling with shoes and, in one extreme case, a child who almost choked herself on her own ribbon. The teachers had to work hard to sustain the observed levels of attention. They persistently reminded children to attend and frequently directed questions to those whom they though were disengaging.

It might be anticipated that getting ten very young children to share ideas and learn from each other could create certain problems. The vast range of differences in attainment and experience have already been indicated and in this light it could be expected that particular contributions would be entirely familiar or totally strange to many participants. The teacher's task is not only to keep discussion to the matter in hand but to sustain progress in learning and help children to render spontaneous contributions meaningful. Some of the processes involved as teachers attempt to sustain learning under these circumstances are illustrated in the following transcript of a session conducted by Mrs. E. The exchanges are, necessarily, reported at length.

Mrs. E. planned, 'To discuss fat and thin things which is part of the mathematics schemes' with her class of 6-year-olds. As a preliminary to this lesson some of the children had drawn 'fat things'. On the wall was a picture of an elephant labelled 'a fat elephant' and a picture of a snake labelled 'a thin snake'. Other relevant practical work had been completed.

Mrs. E.: Fat. Where is the fat thing in this room?
(Several hands up)
Mrs. E.: Terry?
Terry: The pipes.
Mrs. E.: No. Where is THE fat thing that we have named as being fat? Matthew?
Matthew: That elephant.

(Several other children call out 'elephant'.)

Mrs. E.: Don't call out. Matthew?

Matthew: That elephant.

Mrs. E.: The fat elephant and we said we'd written it even with fat letters, hadn't we? And that it's . . . sh . . . we have done the word *fat* with *big* letters and the word *thin* for the snake with *thin* letters. Yes. And you were asked on this table — put your hands down — to make something *fat*. And I am sitting here and I am looking at (Mrs. E. takes pencil out of a child's hand) this lovely snowman. Or fat man? Or what?

Carl: Snowman.

Mrs. E.: A fat snowman, yes, a fat snowman. All right. So a fat snowman. We said we never see a thin snowman do we.

June: No.

Mrs. E.: And this is a fat what?

(Mrs. E. picks up Robert's picture.)

Robert: A fat cat.

Mrs. E.: It looks like a fat man to me. I thought he was very nice. I really did think he was very good. Yes, that's a lovely one.

Robert: Terry drawed that in the middle (not clear what Robert is referring to.)

Mrs. E.: Oh all right well we didn't really know about that. Now this was quite a good idea but I wonder who knows what it is (holding a picture up) Andrew?

Andrew: A fat square.

Mrs. E.: It looks like a fat square, doesn't it? You can't really have fat squares if you think about it. Rachel?

Rachel: A fat book.

Mrs. E.: It is a fat book. I think perhaps you knew. Sit down. But there is nothing to show that it's a book, is there? What do you usually have on the front of a book? (Mrs. E. picks up a picture and various children shout out, 'a picture'). A picture or at the very least you have the name of the book, don't you, and all you are going to do is make it into a fat square which in fact you cannot have. Tell me about a square. What is a square?

Simon: It's got four sides.

This is followed by a long exchange about squares, their number of corners and sides and comparison between squares and rectangles.

Mrs. E.: Rectangles can have very, very long sides and very, very short sides. Right. We have digressed a little. Yes? (to Robert who has his hand up.)

Robert: And they are thin and fat.

Mrs. E.: You can have thin and fat rectangles. Yes you can. Well done. What else, Nicki, can you have that's thin? Let's talk about thin things for a minute. We've got a thin snake up there. Now see if you can think up some lovely things for thin.

(Various children call out.)

Mrs. E.: No, put your hands up ... Put your hands up. Rachel?

Rachel: A thin pipe.

Mrs. E.: A thin pipe. What do you mean by a thin pipe (The child points to pipes round the wall.) The ones that you sit on (Rachel nods) the ones that go round the room that bring the hot water? (Rachel nods) Yes, yes, you can have some very thin pipes. In fact we've got a fat pipe and a thin pipe. By just turning your heads you can see a fat pipe and a thin pipe. Terry put your pencil down and pay attention David, something thin.

David: A snake.

Mrs. E.: A snake yes. Terri? (Terri had her hand up)

Mrs. E.: A thin pencil.

Mrs. E.: A *thin* pencil ... Right a thin pencil as opposed to a *fat*, thick pencil. Um. Terry (Terry had his hand up)

Terry: The door.

Mrs. E.: The door? Now where is the door thin? Come and show me.

(Terry walks over to the door.)

Terry: Inside and outside.

Mrs. E.: Show me. Open the door and show me.

Terry: There to there (pointing to the depth of the door).

Mrs. E.: There to there yes. (Teacher points to the thickness as Terry had) ... I want to say 'thickness' but I musn't say that. The width between there and there.

Damon: A triangle!
Mrs. E.: It's not a triangle.
Felicity: A rectangle.

This a followed by another diversion to point out examples of rectangles. The lesson continues with the children pointing out further examples of thin things including a flower stalk, the hand of a clock, the edge of paper, the teacher and the legs of a chair. And continues:
Mrs. E.: Graham?
Graham: Words.
Mrs. E.: What do you mean words?
(Graham points to the writing on the blackboard.)
(Mrs. E. goes over to the blackboard.)
Mrs. E.: Do you mean the actual marks that you make on here? The actual pieces? The actual writing is thin?
(Graham nods.)
Mrs. E.: Yes, I mean that's a word.
(Mrs. E. points to 'Eskimo') but that's not thin. But the actual marks that we make are quite thin. Yes? (To a child with his hand up.)
Child: The spaces.
Mrs. E.: What? The spaces? What, between the words.
(The child says something indistinct.)
Mrs. E.: Mmm. Not really.
(The lesson continues with other examples.)

The whole session from which the above transcript was taken lasted only ten minutes. It exhibited a number of characteristics common to the six oral lessons recorded. The most striking features are the contradictions within the session. For example, by their constant showing of hands the children showed a degree of interest but the teacher had to be persistently concerned with maintaining attention. Perhaps the interest was less in the specific content than in getting a turn? A further contradiction seemed evident in that whilst at some points the teacher seemed determined to stick to a plan (note the very specific opening gambit) she allowed two very large detours. Perhaps these were less allowed than noticed too late? Furthermore, whilst she occasionally took considerable trouble to get children to clarify what they meant by their vague utterances, she was just as likely to dismiss an idea or elaborate it herself. Perhaps her strategy of getting children to clarify meaning conflicted with a strategy to retain momentum or keep to a time constraint. Finally, whilst her aim was to 'discuss' the topic in hand there was almost no discussion. Children's contributions were never more than a short phrase and were always under the instigation of the teacher. There was no child-to-child talk. Whilst many children made some response, contributions from the class were dominated by only a small fraction of the pupils. Perhaps by 'discuss' the teacher meant 'practice exemplifications'.

These questions will be discussed more fully when we turn to the teachers' perspectives. For the moment it is interesting to shift attention from the processes to the content of the discussion. Why was the discussion about word usage? And, given that it was, why was it so laboured. Why, for example, was Mrs. E. so unwilling to refer to the 'thickness of the door'? Of courses, we can note her conundrum in that the child had used that dimension as an example of 'thin' but there is no inconsistency here in either technical or common usage.

The discussions we saw all had this rather strained quality about them as teachers appeared to struggle against the vernacular and towards some image of consistent use. The exchanges were rich in potential for broadening children's experience of English usage and much was there to be learned from the pointing and showing accompanying the talk. But much was apparently obscured by the teachers' being caught in a tension between English usage and a notional 'mathematics usage' — unnecessarily so for terms which have no specific technical meaning. The teachers had apparently been misled by their reading of the schemes which emphasize the importance of the self-conscious use of terms. Here we see the schemes acting as an important determinant of the quality of mathematical experience because, to a large degree, they influence not only the content but the manner of its treatment. Notwithstanding the teachers' reservations about schemes, it seemed to us that they played a significant role in the mathematics activities we observed.

Children's Investigations

We saw plenty of practical work associated with the pencil and paper activities of the workcards. In this context the word 'practical' refers to the use of concrete props — such as interlocking cubes — used to complete calculations indicated on the cards.

We also saw workcards in operation in which the essence was

practical work. For example, on cards dealing with capacity, children were asked to find out how many cups of water it took to fill a basin or a variety of other receptacles. These cards were presumably intended to enlarge children's experience of quantity. The work in practice never seemed to be quite what the cards intended.

In one instance, for example, a group of four 7-year-old girls had been set the task, from their mathematics workbooks, of measuring aspects of the classroom using a self-made measuring strip marked out in cubits. They made a card strip one cubit long using the biggest girl's forearm. They used the strip to produce seven other cubit strips. This they did very roughly: all the cubits were of different length. Following the workbook instructions they then used sticky tape to hold the strips together. Sometimes the strips were butt-ended, sometimes they overlapped, sometimes the ends were separated by up to half an inch. The technical difficulties of producing the composite strip taxed the girls. It broke several times. The sticky tape seemed to adhere to everything and especially to their socks. It took two mathematics sessions to make the eight cubit strip. On the third session the strip was used in a rough and ready way to measure the items designated in the book. Whilst the girls worked persistently on this card it was difficult to see what they had got out of spending so much time in this way. They struggled to meet the demands of the task as set but the mathematics content seemed very limited compared to the time taken to produce the necessary equipment. Of course, it is easy to point out that with more adult support the task could have been completed more expeditiously, achieving a more rapid focus on the critical element of the concrete experience of length. Equally obviously the teacher cannot be everywhere at once.

In a contrasting example of practical work we observed some 5-year-olds, each working independently and each ostensibly comparing, following the teacher's instructions, which of a selection of containers held more or less water than a plastic bottle. They had been told to fill the bottle and pour the contents into each of the other containers in turn in order to answer the set question. This problem seemed to be rapidly ignored. The children did a great deal of filling and pouring with the containers. They evinced rapt attention to these processes. They seemed to pay special attention to overflowing and promptly avoided, it seemed to us, cases in which pouring did not fill to overflow. They stacked the containers arranging them to get cascades when the top one was overfilled. They repeated filling sequences over and over again. The children were not passive in this work. They appeared to be setting themselves problems. Some of the cascades evidently did not work to their satisfaction. The position of containers was consequently shifted to get more acceptable effects. Whilst these children had worked extensively with water on previous occasions their persistent attention to this activity, indicated, it seemed to us, that they felt there was much of interest to be learned here. This view was shared by the teacher who said, 'Their eyes never glaze over when they set the problems.'

Whilst this teacher was very happy with the activities of her children, these activities had come about because the youngsters had ignored a problem set by her. We saw only one case in which the teacher deliberately let the children define and develop the problem from scratch. She had a motive which set some structure on the events. The motive necessitated that she allowed a great deal of flexibility.

Mrs. A. had received a new child, Ian, into her class. On the face of it, Ian was well advanced in the mathematics scheme but his performance led Mrs. A. to believe that he did not understand much of what he had covered. She felt he needed a lot more practical experience and had attempted to engineer this by getting him to rework some exercises. Ian had resisted this and Mrs. A. concluded that he felt it was beneath his dignity to go back on the scheme. In consequence she arranged for Ian to 'help Ben' who was a year younger. Ben had been carefully chosen for several reasons. He was judged to need practical experience — especially with the concept of weight. He was as affable as Ian was insular, as adventurous as Ian was defended. They got on well together. By getting them to work with a beam balance Mrs. A. predicted that each would learn something — and so would she, 'I'm not sure I am right about Ian. He could be better than I imagine. We'll see.'

In the sessions previous to the one to be reported, Ian had been working with a balance to compare objects to find which was the lighter and which the heavier. He had been successful with this work in that he had interpreted the balance correctly. Mrs. A.'s concern was whether he understood the notions of heavier and lighter and whether he could be got to think more productively with these concepts. She gave Ian and Ben a beam balance and a deliberately familiar starter, 'Find heavier and lighter things and see how many small things weigh the same as the large blocks.' The boys had access to a huge box of assorted wooden blocks to select from.

Ian appeared to take the lead. He began to tackle the work as if it were a repeat of his previous exercises. He got out his workbook and started to write out appropriate sentences and he encouraged Ben to do likewise. Ben watched but did not write. Ian chose the objects, loaded the balance and wrote down the results. After two comparisons he sent Ben off to choose two shapes.

Ben brought back a cube and a cylinder. He put them in the pans. The cube was heavier. Ian swapped the objects round. Ben looked very surprised at the new imbalance. He swapped the objects back and then back again. He continued to look puzzled. Ian wrote the result down.

The teacher came over with two large pine cones. (She later explained that she wanted to get Ian out of a rut.)

Ben put a cone in each pan and propped one pan up with his finger. He said, with a huge grin on his face, 'This one's gone heavier.' Ian said, 'No, it's gone lighter.' Ben looked puzzled again and then agreed. Ian wrote the result down. Whilst Ian wrote, Ben fiddled with the balance.

The teacher brought over a tray of objects — mainly comprising wooden beads of various sizes. She said later that she had done this because they seemed to be messing with the balance rather than using the objects available.

A discussion ensued between the two boys about the objects and their relative weights and sizes. Both seemed to use visual appearance to judge weight. In cases of dispute they used their hands to balance the respective weights.

Ian started to write some results down. But Ben began to make a balance by putting a long strip of wood across a block. He put some beads on the ends but they either fell off or nothing happened. Ian began to copy this approach. He immediately used a smaller pivot block but had no more success than Ben. Ben, in his turn announced that he needed a 'longer piece' but evidently could not find one. Instead, he copied Ian's idea of using a smaller pivot block.

Ian tried a cylinder as pivot but he could not stop it rolling long enough to balance anything. Ben repeated the suggestion of using a longer beam. In response, Ian began to build a longer beam in the fashion shown in figure 5.

He persevered with this design for about five minutes but it collapsed at each attempt. He gave up and went back to the beam balance. Ben however, found a triangular prism to use as a pivot but just as Ian was attracted to try this the bell went for play time.

In this session we saw children constantly concentrating and setting themselves problems. We saw them provoking each other's thinking and sharing ideas. There was no capitulation to authority — whether it be the teacher's or their partner's. Perhaps we saw Ian weaned ever so slightly from his notebook and Ben moved a little

Figure 5: Ian's attempt to extend a balance beam

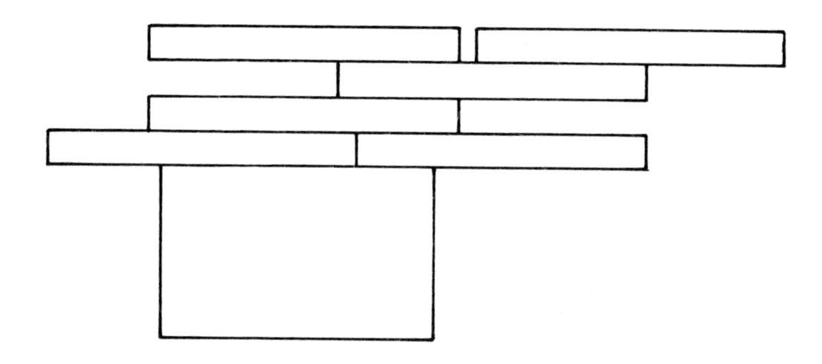

towards directed exploration. We saw very subtle processes of teacher-pupil interaction manifest not in words but in the judicious provision of materials. This action arose from the insightful observation of the teacher who, necessarily, was running the rest of her class as well as fostering the activities of Ian and Ben.

The activity was designed because the teacher asked herself a question about a child rather than asking children questions about mathematics. In her view, and in the light of distracted observations of the session, she concluded tentatively that Ian did have an appropriate grasp of the concept of weight but was clearly in need, as was Ben, of a lot more experience with materials. To the adult, the device in figure 5 simply cannot work. This was not at all obvious to Ian or Ben. As Mrs. A. put it, 'Children and materials need an awful lot of mixing.'

Summary

From our perspective the major, immediate impressions of classroom mathematics were of a lot of extremely industrious children responding to efficient and business-like teachers and engaged, in the main, in exercises drawn from commercial schemes. Whilst practical work was evident, it played, by and large, only a supporting role to paper and pencil work. The children very rarely tackled real problems in their mathematics or applied their thoughts to anything other than the procedures demanded by the worksheets.

In these respects the classrooms observed were very much like the classrooms reported in many studies on both sides of the Atlantic (Bennett *et al*, 1984; Sirotnik, 1983). The teachers' achievements were substantial but the limitations, in terms of ideal practice, as described for example in the Cockcroft report, seemed obvious. But how did it look from the participants' point of view?

Chapter 5

The Children's Response

How did the children respond to the mathematics activities in the classroom? This is an important question because, following Doyle, we might anticipate that the children's responses to their work might play a part in informing the teachers' behaviour in planning and setting that work. To understand the teachers we have to grasp the children's response. Were the children happy and successful with their work? What work practices did they deploy? What skills did they evince? What feedback was potentially available to the teachers as their children set about the tasks assigned?

We obtained data on these matters in a variety of ways. Before our periods of observation in the classrooms we identified, with the teachers, the work to be covered during our stay. We then designed a short test on this curriculum content. This was to ascertain whether the children had appropriate prerequisite skills to tackle the content and their degree of familiarity with it. We observed target children (see chapter 1) closely whilst work was assigned, and throughout the allocated mathematics time. We conducted post-task interviews with the children to try to understand the immediate effect of the tasks set. It should perhaps be emphasized that we were not very interested in how much the children learned. We were to be in each classroom for only three weeks and our experience with young children led us to expect very little progress over this time scale. Furthermore, the teachers had indicated that they took the approach that their task was to build a firm foundation based on solid and safe acquisitions of concepts by means of plenty of practice — an approach also emphasized in the scheme they used. Indeed, contrary to rapid progress we might have expected to see many instances of short-term regression. As Mrs. A. observed (see chapter 2) children in this age range seem occasionally to lose learning in one area whilst they take on something new and substantial in another. Consequently, rather than quantifying the progress made, we were more interested in identifying the processes by which the teachers and children understood each other.

To this end we had complete sets of data, including preliminary tests, on-task observations and follow-through interviews on 113 tasks covering the work of fifty-three different children, that is 26 per cent of all the children in the classrooms observed. It is these data that are used in the following discussion. Of course we could (and did) collect our data in a much more leisurely way than could the class teachers. This put us in a better position to understand the selections teachers made from the data potentially available to them.

We took as our starting point, in the collection of data from the children, the teachers' descriptions of the work they intended to cover during our period in their classrooms. These descriptions identified a set of concepts and the set of tasks to be used to develop these ideas and skills. In designing a test to assess children's knowledge and attainments relevant to this anticipated content we had to probe skills and ideas both simpler and more complex than those intended to be covered. This was necessary to assess whether the children were ready to begin these sessions or were well in advance of them and hence to put us in a better position to understand their response to the work set.

For example, we met Mrs. B.'s reception class a few weeks into their first term in school. They had done some basic work on sets, relations and counting and Mrs. B. anticipated that the group would begin work on the cardinality of small numbers, on identifying written number symbols, on making and identifying a set of a given number of objects and on relating written symbols to quantities — all in the range nought to five. (It should be said that this was the work likely to be covered by the range of her children and not by any particular child). One of the workcards the children were expected to use is shown in figure 6.

In order to assess the children's attainments with respect to these concepts we designed a small test which started with simple ideas but which, for those who showed a capacity to cope, went much beyond these. For example, we checked the children's ability to count three similar objects arranged in a line, a circle and a heap. We asked them to show us how far they could count (that is in the sense of simply reciting numbers), we asked if they could make us a set of four pencils, if they could identify the number symbols to five and make a set containing the number of toys shown on a card exhibiting a number symbol. But our test went beyond this, first to the range of

Figure 6: A standard work card

mark each set of 2

mark each set of 3

mark each set of 4

mark each set of 5

bigger numbers, including numbers over ten where the child seemed confident with smaller numbers, and subsequently to ask them if they could make quantities 'one more' and 'one less', to present them with adding up problems with small numbers and, finally, to present them with verbal sums of the type, 'If you had two sweets in one hand and two sweets in your other hand, how many sweets would you have altogether.' Appropriate tests were designed for each class we studied.

The tests were given individually to the children in brief conversations spread over a week. They frequently involved the use of pictures and concrete objects arranged in the form which the children's mathematics schemes exhibited. Practical, oral and written responses were requested as appropriate to the concept or procedure being assessed and the child's responses were recorded in a pre-prepared test booklet. The tests were given individually to the target children and only after we had established some familiarity through chatting over several days. The complete test was not given to every child. If a child could not cope with the easy questions his ability was probed on the next level and, if unsuccessful, the test was stopped. In the test briefly described above, for example, Darren, aged 4, was unable to count (in recitation) beyond two and could not say how many in a set even with only three objects. He was not troubled with much more of the test.

As might be anticipated, we found huge differences between the children even at age four. In Mrs. B.'s class, for example, Michael (aged 4;2) could recite numbers to six when asked to count but could do no more. In the same group, Andrew (4;3) could recite to fifteen, count up to eight objects in a set however arranged, consistently identify the number symbols to ten, write out these number symbols, understand 'one more than' and 'one less than' and could do simple adding and subtraction sums (including $2 + 2$ and $5 - 2$) if the objects were present and a concrete context were supplied. These observed differences proved crucial in understanding the differential response of Michael and Andrew to subsequent exercises.

With reference to the exercises used, it was clear that each required a great deal more than mathematical conceptual understanding for their completion. To succeed on the exercise shown in figure 6, for example, the child would, if left to his own devices, have to be able to read the instructions, understand what was meant by 'mark', understand what would be acceptable as a 'mark', understand that each box was a set and be able to wield a pencil or crayon. Additionally, he would have to understand the symbols at least to the degree of being able to separate elements for counting purposes. The three overlapping buffalo or antelope, for example, are only visible to those who see the symbols as buffalo or antelope. They could be blotches or a single blotch. They could show sets of six horns, a number of legs — and so on. Given all these facilities, the child who also had a grasp of the relevant number concept would be at an advantage with such a task. In this light we have found it useful to make a working distinction between conceptual and procedural competence.

In principle, to complete a card with due mathematical integrity, a child would need to have some grasp of the appropriate mathematics concept (conceptual competence) and to know what to do on a particular card (procedural competence). In figure 6, row two, for example, a child who had a grasp of the 'threeness of three' would not necessarily be able to perform successfully. In addition to his number concept he would need all the task specific comprehension described above.

Procedural Learning

In recognition of the demands of the worksheet, the teachers — as we saw in chapter 4 — always spent plenty of time ensuring that their children understood the particular requirements of the task in question. In this practice they were encouraged by the teachers' handbooks associated with their schemes. It seemed a very necessary practice since the demands of the cards were not, in a predictable way, related to the anticipated attainments of the pupils. Figure 7, for example, shows a card thought suitable for a child in the first term at school whilst figure 8 shows a card considered appropriate to term six.

Notwithstanding the demands of the cards, the teachers' copious attention to the production details was frequently not necessary. In 51 per cent of the cases observed the children knew beforehand precisely what to do before they were told. For example, when six year old Sasha was shown the card illustrated in figure 9, she commented, 'I have to put the number and then I have to draw the number of pictures it says to draw' — an interpretation she then illustrated by writing '13' and drawing thirteen triangles.

Sasha appeared to give no sign of impatience, resentment or frustration when her teacher showed her and her group how to do this card. Very occasionally children did tell their teacher that they knew what to do already but it did not deflect the teachers from their strategy:

Figure 7: Workcard for term one

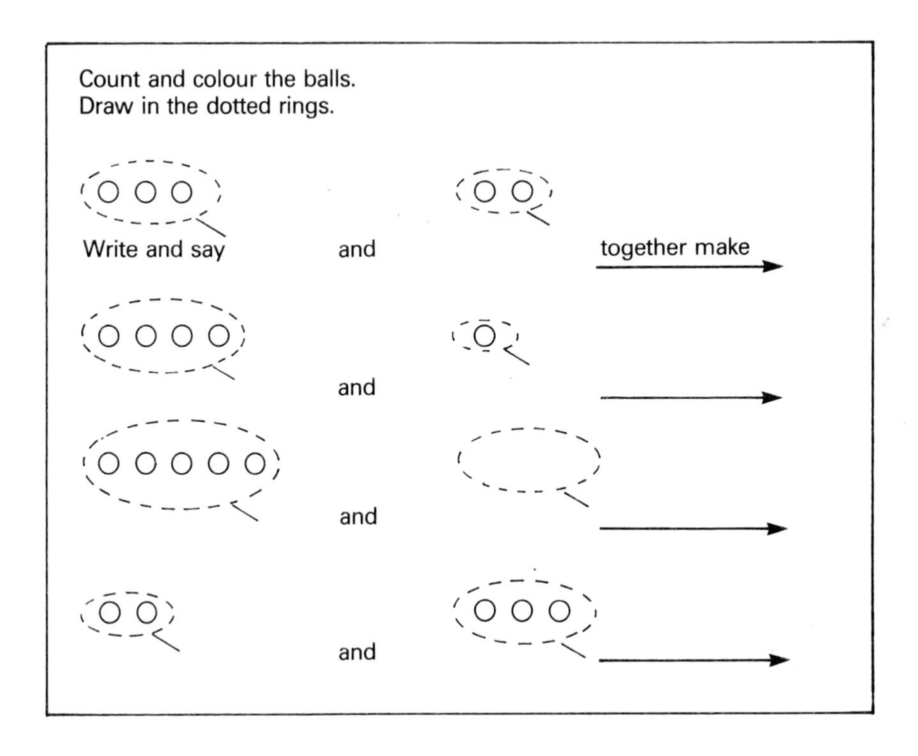

Count and colour the balls.
Draw in the dotted rings.
Write and say
and
together make
and
and
and

Figure 8: Workcard for term six

2 + 1 = □	0 + 4 = □
3 + 2 = □	4 + 2 = □
2 + 3 = □	5 + 0 = □

Figure 9: Sasha's workcard

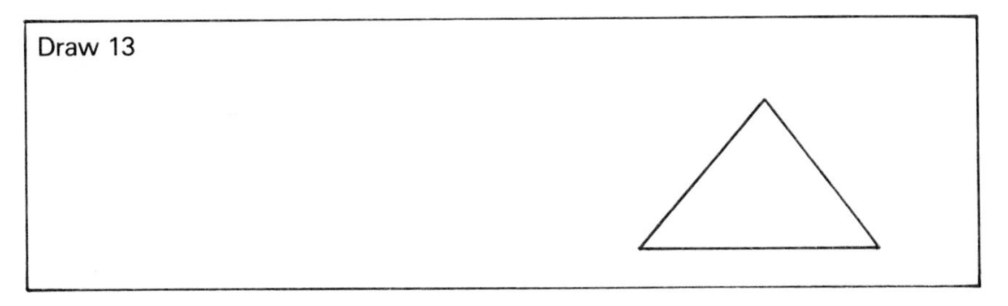

Mrs. F.: (approaches Joanna) Now Joanna . . .
Joanna: I know what to do . . .
Mrs. F.: Well let's go all the way through and make sure.

And Mrs. E. reacted as follows to Terry's impatient fidgetting in a group:

Mrs. E.: Wait a minute Terry and I'll tell you what to do.
Terry: I know what to do.
Mrs. E.: (pointing to the top of the page) What does it say at the top here?
Terry: 'How many?'
Mrs. E.: So what are you going to have to do?
Terry: Count how many and put the number in the box.
Mrs. E.: Excellent. Well done. Off you go.

Such announcements or other signals of competence were rare. Mostly, the children appeared very content to be assured of what they already knew. And the teachers could hardly be expected to risk that they knew. Such a strategy would have been wrong in 49 per cent of these cases observed and would have led to certain frustration and potential chaos as children would have had to have waited for clarification.

The teachers' copious attention to detail in teaching procedures frequently paid dividends. They succeeded in 40 per cent of all cases seen in teaching a child a procedure they did not already know. (This amounts to only 20 per cent of all cases because in half the cases of teaching, the children already knew the required procedures). For example, prior to a task on the sixness of six, 4-year-old Thomas could count six items but was unable to write '6' or recognize it. After a five minute session, however, he proved able to write it readily and could get the appropriate number of items when shown the symbol. Whilst Thomas's task was, perhaps relatively straightforward, other children mastered more complex procedures. For example, 5-year-old

Figure 10: Aaron and Tony's task

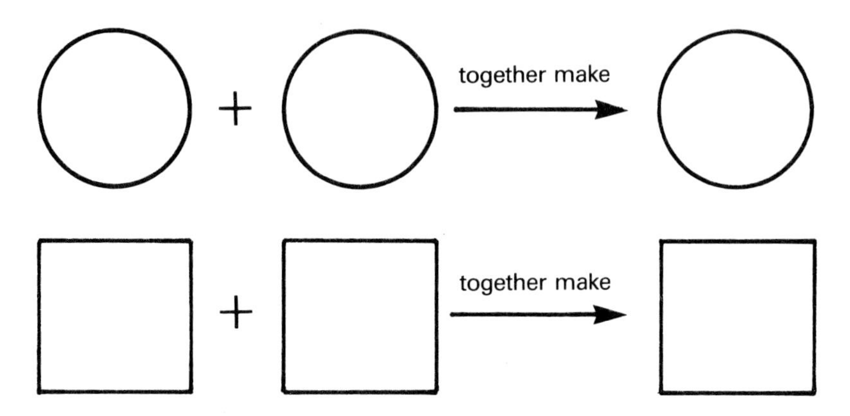

Aaron and Tony learned how to cope with the worksheet shown in figure 10.

As Tony explained:

> You had to put as many as you want in there (pointing to the first circle) and some in there (pointing to the second circle) and then put them all in there (pointing to the third circle) and count them and then you had to put the numbers in there (pointing to each square in turn).

Pre-task interviews had shown that Aaron and Tony had mathematical skills appropriate to this task. They could both add up by counting-on. What they learned from this session was how to do the card.

Despite the teachers' attention to detail however, they failed more frequently than they succeeded to teach children pertinent procedures. Of all the children who could not do a card before the teacher started her instructions, 60 per cent still could not do so twenty minutes afterwards. (This amounts to only 30 per cent of all cases seen because, as we have observed before, 50 per cent of children could already do the procedures.)

Mrs. B., for example took a group of four year olds through the procedure appropriate to the card shown in figure 11. It should be noted that the children had previously done a lot of practical work on adding up in this number range.

> *Mrs. B.*: Now here we've got a ring (circling the set) and a line going across it to cut it in two bits. Now then, in that side of the ring (points to the left) how many have we got?

Figure 11: Mrs. B.'s lesson on addition

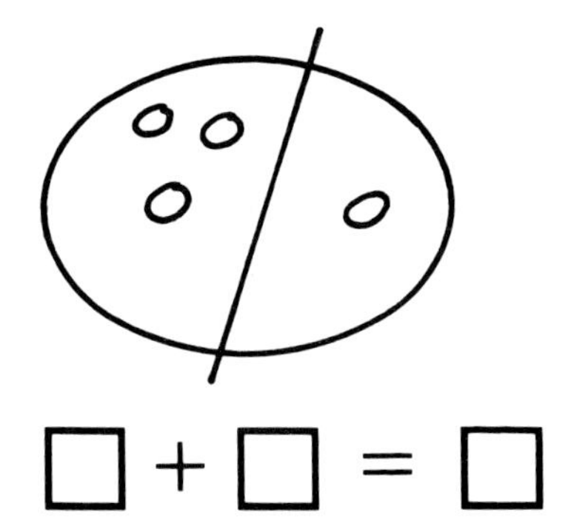

Helen: Three.
Thomas: Four.
Mrs. B.: How many, Thomas (surprise in her voice) On that side of the ring how many have we got (points to the right)
Thomas: One.
Mrs. B.: Right. So there are three there and one there. Now then. Underneath are two boxes and the first one wants to know how many in that part of the ring (points left) and so what would we put there?
Paul: Three.
Mrs. B.: That's right. And the next box wants to know how many in that part of the ring (points right). So what would we write in that box?
Helen: One.
Mrs. B.: And the last box wants to know how many there are altogether (circles set).
Helen: Four.
Mrs. B.: Four. So we would write four in the last box wouldn't we? And that's saying (pointing) three add one makes four. And when we write it down we say, three add one makes four. Know what to do?

Only Helen nodded and Mrs. B. went through two more examples before asking the children to do their exercise. Paul and Thomas evinced difficulty with this task and subsequently Thomas said to the interviewer, 'There are too many boxes and I don't know what to put in them.' Pre-task interviews had suggested that despite their previous work neither of these children had the necessary pre-requisite skills to do this task. Neither could add up in any form with concrete objects let alone do two stage sums involving making the sum up before solving it. In this and other instances of failure the children's problems were met with repeated demonstrations of the routine rather than an effort to diagnose any underlying problem.

A very similar case was recorded in chapter 4 where, it will be recalled, Mrs. D. set out to teach her children how to do a form of subtraction sum from a partitioned set. In that instance, repeated demonstrations did, at first sight, appear to be successful in getting children to follow the desired routine. The children's behaviour was praised and, on the next day, the teacher moved on to a new task. Immediate post-task interviews had shown, however, that the children had not grasped Mrs. D.'s lesson.

Clues to the teachers' failings were not immediately available to them. Whilst children often showed initial confusion, detailed instruction frequently had them doing as they were told. Only in follow-up interviews, with no teaching structure and no friends or other props available were the children's shortcomings revealed. The teachers were not seen to indulge in such interviews.

As we observed in chapter 4, most children ended up doing as they had been told — at least in the short term — and this must have been very reinforcing for the teacher. The fact that half of the children did not need telling could not be taken for granted by the teacher faced by particular instances. And the children, in the main, appeared happy to be told. The fact that almost a third of the children did not grasp the procedure was hidden because most of that number passed muster on the day. If the teacher saw a child in difficulty, she re-taught and that seemed to do the trick. Additional post-task follow-up interviews must have seemed unnecessary because the children were happy with their completed cards. The teachers might have considered themselves busy enough without doing unnecessary work. There were always a few children who could evidently not do the work even after being shown several times and these seemed to attract most of the teacher's attention.

Conceptual Learning

Whilst the teachers emphasized procedures to the children, their expressed aims laid a heavy stress on the development of conceptual understanding. Important aims articulated included:

To try to deepen their understanding of the concept of solid shape.

To develop their understanding of addition by moving from concrete to abstract examples.

To introduce and develop the relationship between addition and subtraction.

As indicated earlier, we anticipated that the growth of understanding of mathematical concepts would be a gradual process in which the teachers built on considerable amounts of practice and consolidation. This expectation was confirmed. In 66 per cent of tasks observed the children already had a good grasp of the concept at the heart of their tasks. A week before his first lesson on subtraction, 5-year-old Paul explained a problem to us thus, 'Six take away six is nothing. If I had six buns and someone ate them all up I'd have no buns left would I?'

Many of the children demonstrated in interviews that they understood the structure of a problem by showing other ways in which to do it. Six-year-old Joanna was taught to draw round Stern rods to complete a series of addition sums. Thus for '4 + 3' she meticulously — but with great difficulty because she could not hold the rods still — produced the following:

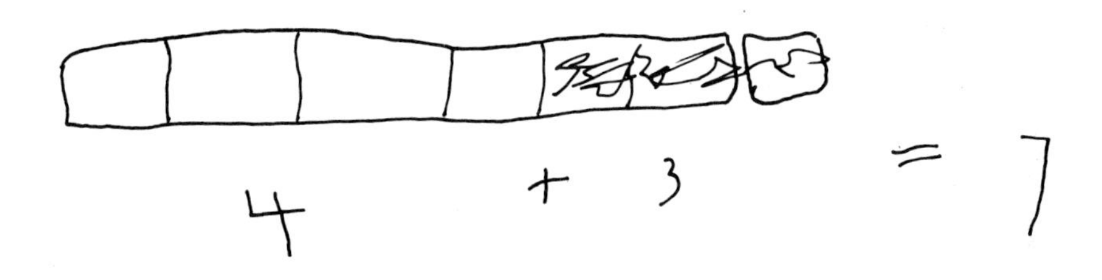

When we discussed this problem with her she told us that it was easier to do them in your head and if the numbers got bigger you could always use your fingers.

Similarly, 6-year-old Darren was seen using a standard procedure with Unifix cubes to complete addition sums with a limit of nine whilst in a previous interview he had exhibited a wide range of approaches to such problems:

Interviewer: Try this. 5 + □ = 6.
Darren: Five to make six (immediately writes '1' in the box).
Interviewer: Okay. 2 + □ = 10.
Darren: Two to make ten. (Holds up 10 fingers. Puts down two, counts the rest and puts '8' in the box).
Interviewer: Okay. 5 − □ = 2.
Darren: Five take away to make two (immediately writes '3' in the box).
Interviewer: Now this, 11 − □ = 5.
Darren: Oh dear I haven't got enough fingers. (He puts a pencil on the desk and his fingers alongside it). I've got eleven fingers now. (Puts down five fingers and nods serially at the rest — presumably counting them. Puts '6' in the box).
Interviewer: Try this, 10 + □ = 16.
Darren: (Writes '10' in the box. Then) No! That would make 20. Oh, it's six!

It seemed that Darren was well in advance of the tasks his teacher set him. He appeared to have a number of strategies for doing both addition and subtraction problems including direct answering, using fingers, using props and checking answers. Left to his own devices he used the technique that he felt he needed, resorting to props only when he could not directly answer. Told what to do he did as he was told.

We are not suggesting that because, in two out of every three tasks observed the child grasped the concept entailed in their tasks before they did them, that they were therefore wasting their time. The teachers were committed to a high rate of consolidation. The schemes they used suggested that this was important. It should also be stressed that consolidation is much more than mere 'hammering home' or 'overlearning'. Several research studies have shown that children invent new ways of doing things as they practise familiar routines (see Resnick and Ford, 1981, for a review of some of these studies). Consolidation work can thus lead to the very deepening of understanding to which the teachers aspired.

From the teachers' point of view the children were happy and successful in their work and were making progress through the schemes. The children revelled in the praise their performances brought them. This scenario proved remarkably similar to that observed in our previous study of infant mathematics (Bennett *et al*, 1984). If we combine the instances in which the children knew the procedure and those in which they knew the concepts involved for a given task we can conclude that in 40 per cent of all tasks the children knew and understood exactly what to do before the task was set. This level of practice is actually lower than the 50 per cent we saw in our previous study and might suggest a faster rate of progress through the work.

In almost all the remaining 34 per cent of tasks seen the children did not understand the concepts involved although they could frequently do the routines to the satisfaction of their teachers. Leah provides a typical example of a child in this situation. She had attended several sessions on the relationship between addition and subtraction. In one of these sessions the 7-year-old was observed correctly formulating sums from sets. For example, beside the figure:

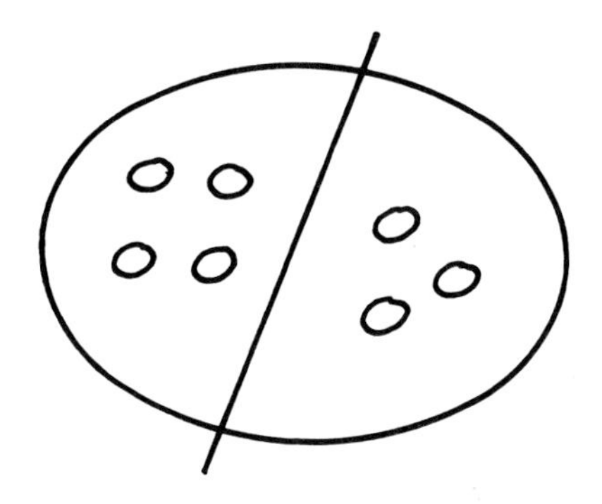

she had written:

$$4 + 3 = 7$$
$$3 + 4 = 7$$
$$7 - 4 = 3$$
$$7 - 3 = 4$$

By the end of this session the teacher had concluded that Leah had mastered the underlying concepts involved. In a post-task interview Leah was offered a diagram showing one object and two objects on either side of the partition line in a set. She quickly jotted down, $1 + 2 = 3$; $2 + 1 = 3$; $3 - 1 = 2$; $3 - 2 = 1$ beside the set. However, when asked to make up a problem about her first number statement she said, 'I went into a shop and bought three tangerines and one banana and that makes two altogether.' and, 'I bought a pencil for 3p and a clock for 2p and that makes 1p.' This did not surprise us because in a previous post-task interview Leah had shown no proficiency with addition or subtraction with numbers or with word problems. On this basis we doubted whether Leah had grasped the new concepts from her recent work with Mrs. D. However, Mrs. D., on the evidence of an error free work sheet concluded otherwise.

The difference between our views and the teacher's interpretation has its basis in the evidence we each had to go on. We both saw a happy child working hard to produce an error free sheet. We had the additional evidence obtained through interviews in which we did not attempt to teach or correct the child. We let her demonstrate her comprehension — or lack of comprehension as it turned out in Leah's case — to us. The teacher did not use this technique. The teachers' perspective on this matter is illuminated in the next chapter.

Once again we have to stress that we cannot conclude from these observations that Leah and the children who, like Leah were doing work they did not understand, were wasting their time. Understanding can certainly come from work which is beyond one's current comprehension. Leah and most of her ilk were having sufficient procedural success to sustain motivation. Given this, they were likely to continue to work on tasks demanding relevant concepts. And in making these observations it is clear that the distinction between 'procedural' and 'conceptual' competence, whilst useful to make a point, is an uneasy one. The only way we can know a child has a concept is to give them something to do that involves the use of that concept. But such tasks can never be 'pure', they must always involve more than the concept itself. Indeed there can be no useful distinction between having a concept and being able to use it on a real life — probably messy — problem. And there is probably no royal route to gaining the facility to apply concepts. Sometimes we might expect understanding to come before use but it seems at least as valid to expect understanding to come from use. Whether activities are pertinent to the acquisition of a concept is a matter for judgment in a particular case. Since almost all the activities we saw assigned came from commercial schemes it seems reasonable to conclude that the teachers had passed that judgment over to the experts who design these materials.

Certainly, the systematic use of these materials was associated, by the teachers, with broad progress on the part of their children. This is a view consistent with our own previous research findings (*ibid*). In that study, sixteen infant teachers had used commercial schemes as the main plank of their mathematics curriculum and tests at the end of the year showed that high attainers had retained in excess of 90 per cent of the material covered whilst low attainers retained 75 per cent on average.

In the present project, some individual advances were quite striking and must have sustained the teachers faith in their approach. In Mrs. G.'s class, for example, Gary (7;2) and Kerry (6;3) had both received class lessons on aspects of the concept of place value. In pre-observation interviews both had shown a patchy grasp of this matter. They were able to recognize a selection of numbers in their symbolic form up to 100 and when given a pair of numbers (for example sixty-three and thirty-six) consistently knew which was the larger. They were also able to identify the pattern in the series, '4, 14, 24, 34' but they were unable to supply the next figure. They could write any of the numbers up to twenty. However when orally given a series of calculations such as, '10 + 6; or 16 − 6; 10 + 4 or 14 − 4' they had to work out each answer in turn. They had no spontaneous number bonds in this range, nor did they see the similarity between the pairs of problems.

After the first lesson on place value, Kerry was shown the phrase, '1 ten and 5 units' and asked to show it with Unifix cubes. Kerry put out the following and said:

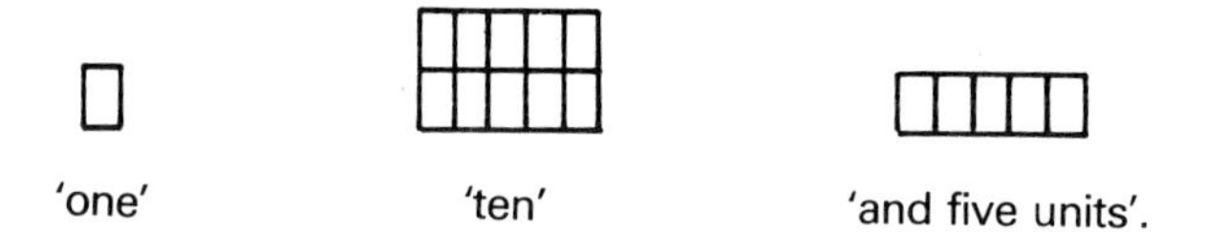

When she was asked, 'How many are there?' she counted the cubes and announced, 'sixteen'. When asked how many the phrase, 'one ten and eight units' was she said 'nine'. Gary's answer to these questions were similar to Kerry's. When asked for the value of 'one ten and five units' and 'one ten and eight units' he said 'six' and 'nine' respectively.

Four days and three lessons later the children were shown a cluster of fourteen dots. By counting, they each ascertained that there were fourteen. They also knew that that meant there was one ten in there but they did not know how many units that left over.

After only one more lesson, in which the teacher discussed place value using the numbers twenty to twenty-nine, both children showed a clear understanding of place value. They could split any numbers under 100 into their component tens and units (the teacher had not covered beyond twenty-nine) and describe their reason for doing so. Gary described 'forty' as 'four sets of ten' and Kerry said that 'thirty-three' was 'three lots of ten — ten, twenty, thirty — and three left over'. All the questions that had defeated them in earlier interviews now seemed trivially easy and they got them right quickly.

These children's successful performances, over a week, on tasks they certainly did not understand, seemed closely associated with their apparently sudden grasp of a notoriously difficult concept. By rewarding their efforts and performances the teachers had kept the children engaged on a series of tasks which, in the end, paid dividends.

It seems then that the teachers assigned tasks from high status sources and were able to observe their children working cheerfully, willingly and to a large degree successfully (in terms of completing procedures) on them. The teachers took a great deal of trouble to spell out the requirements of each task. In 40 per cent of cases this trouble was not necessary because the children knew and understood what to do but they rarely let the teachers know this. In 20 per cent of cases the instruction went home in that the children picked up a new routine and in this sense broadened their experience and competence. In the other cases the instruction had, in the short term, only a transitory effect; it enabled the child to complete the task but not necessarily to either understand it or repeat it without help. In any event the children were busy and evidently happy. These substantial general patterns still left a great deal of scope for individual responses amongst the children. They were by no means robotic.

Fixing the Task

We saw in chapter 4 that despite the teachers' very close instruction, a lot of children had their own ways of doing tasks. Some methods were both mathematically appropriate and genuinely inventive. Others were less so. We noted, for example, that some children were adept at looking back through earlier work to re-discover an answer, others spotted the 'secret of the page' whilst yet others rushed to complete work whilst they could still remember the answers from their teachers' demonstrations. In these cases the children were altering the demands of the tasks set by their teachers.

The teachers were aware of many of these strategies and pointed out several interesting cases. In Mrs. F.'s class, for example, Nicola and Roscoe evinced a tendency to copy. They had evidently started with the odd instance of 'assistance' but had begun to develop the strategy into a wholesale work sharing cooperative. Whenever they were given the same work sheet they attacked alternate calculations and copied the intervening ones from each other. The teacher decided to draw their attention to the limitation of their approach by setting them sheets that looked the same at first glance. Their response is shown below.

Roscoe:		
	T	U
	2	8
+	3	1
	5	9
	T	U
	4	0
+	3	6
	7	3

Nicola:		
	T	U
	1	3
+	4	5
	5	9
	T	U
	4	2
+	3	1
	7	3

They stuck to their work sharing strategy throughout the ten sums on the workcards and at no time realized the differences between them. They were intent on completing their work 'efficiently'.

Mrs. A. pointed out to us a more subtle case of task alteration evinced by Ian whose work we described in chapter 4. Ian had been set a card calling for him to make comparisons between the weight of different objects. He could choose the objects but in each instance he had to complete the sentence, 'The ... is heavier than the ...'. Ian selected various pairs of objects and worked steadily making his comparisons and recording the results accurately. He then chose a yellow cuboid and a green cuboid, put them on the balance and turned to record his result. This seemed to cause him a great problem.

He looked anxious. He stared at the record sheet. He went back and forth from balance to book. Eventually he took both cuboids back to the brick box and chose two other objects. Never again did he choose two objects of the same shape. When Mrs. A. asked him what the problem was he explained, 'There is only space for one word.' Ian, ever willing to be rule bound, had found that he had to record 'yellow cuboid' in a one word space. To avoid the problem he altered his method of selecting objects. Being rule bound takes quite a lot of thought.

In responding to the teachers' tasks then, we have seen that whilst the children's predominant strategy was to follow the instructions, there was a plethora of alternative routes through their work. These are summarized in figure 12.

The Children's General Perceptions of Mathematics

We have recorded how the children responded to their specific mathematics tasks. Almost invariably they worked with evident industry and pleasure on their exercises. We saw only two cases of upset and neither of these had anything to do with the mathematics content of a task. Gavin was upset when he could not do his workcard because, 'I can't draw fish' and 7-year-old Sasha had a little weep when she could not hold a set of Stern rods steadily enough to draw round them to record her answer to a problem. These instances apart, the children's response to their paper work was very positive. They seemed much more distracted during discussion sessions.

When asked if they liked their work they seemed unwilling to enthuse about discussions (although no one said 'no') but were very keen on the written work. Gary liked mathematics, 'even more than football'. And liking mathematics seemed, to some of the children, sufficient reason in itself to do the work. This was well put by Jason:

Interviewer: Why do you think you do these?
Jason: I just like doing them.
Interviewer: Is that why you do them in school?
Jason: Yes, I just like doing them.
Interviewer: Do you think Mrs. D. likes doing them?
Jason: I don't know.
Interviewer: Why do you think she gives you sums?
Jason: Perhaps she knows we like doing them.

Figure 12: Children's routes through a task

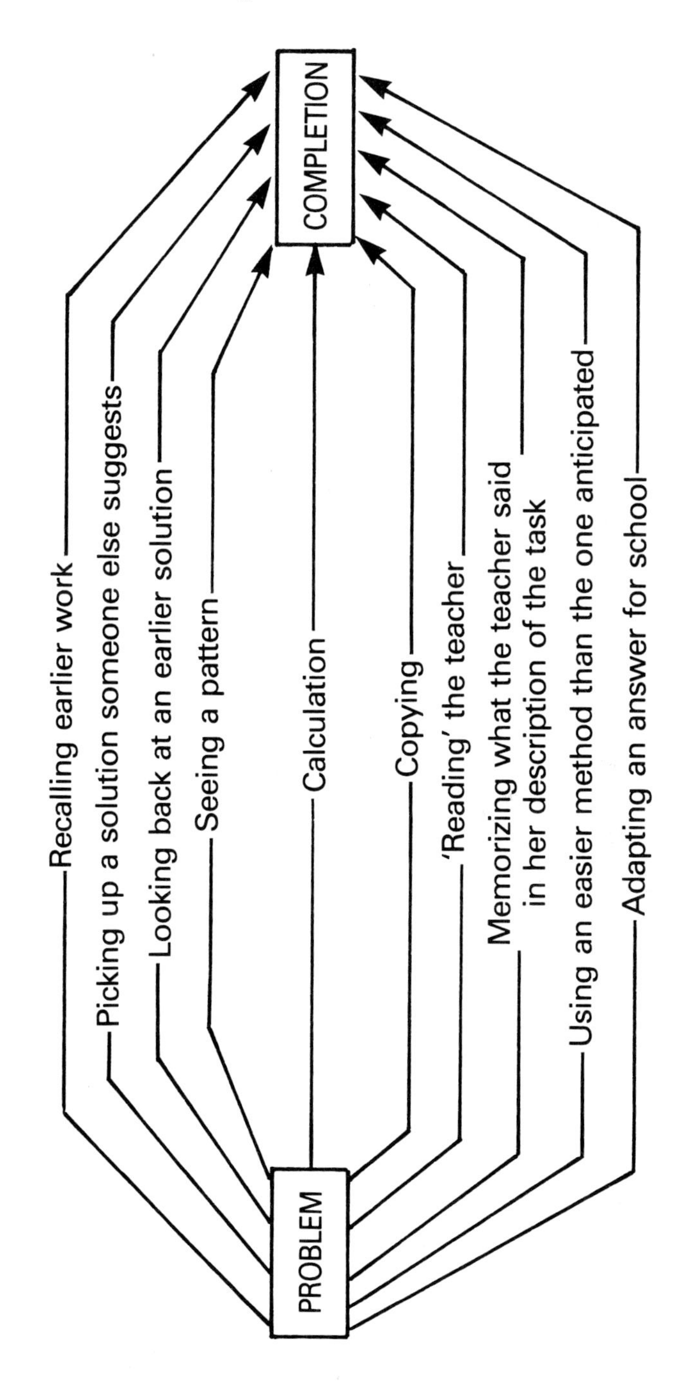

Other children, especially the younger ones, could see more immediate, practical reasons for doing their mathematics exercises:

Interviewer: Why do we do these numbers?
Michelle: So that you can spell things.
Interviewer: Spell things with numbers?
Antoinette: So we can count properly.
Interviewer: What kind of things do we need to count?
Antoinette: You need to count the numbers.
Simon: We need to draw the numbers.
Interviewer: Why do we need to draw the numbers?
Lisa: So we can copy them.
Interviewer: But why do we need the numbers at all?
Lisa: So we can colour them in.

Others could see a longer-term purpose. Ben pointed out that, 'You need numbers to recognize things. I have a chemistry set and there are numbers on each tube and you have to look in the book and it tells you the number to look for on the tube.' And 4-year-old Andrew had no hesitation in explaining that, 'you need numbers in case you are a policeman'. There were no other references to adult life and, indeed, several children cast doubt on whether adults indulged in mathematics activities. Certainly they could recount very few instances of having seen adults doing calculations or measuring. Ben thought for a long time before recalling, '. . . about nine months ago . . . no . . . a year ago, when we moved house, Mummy had to count pegs to put the curtains up with.' Even when they thought of an instance of adults doing something similar to them the correspondence was slight. When Kelly was interviewed after measuring the classroom with an eight cubit tape she at first denied having seen any adult measure anything and then recalled:

Kelly: . . . oh yes. I've seen Daddy measure a lorry.
Interviewer: Did he use a tape?
Kelly: No. It was a chain.
Interviewer: Was it cubits?
Kelly: No it was a chain.

In most cases it seemed that the purpose of doing the mathematics was, to the children, self-evident.

Interviewer: Why did you measure these things?
Kerry: For the book.

Interviewer: Why do you think it's in the book?
Kerry: It's our work.

The majority response to our questions on the purpose of the children's mathematics was that it was 'our work' and that that was sufficient unto the day.

We obtained a more direct view of the children's motivation from comments they made spontaneously — sometimes as they rushed past us to get back to their tables:

I did eight pages today and I got them all right.

I've only got six more pages to go.

I started there and I've done this page, and this page and now I'm on this page and I've only got one, two, three, four, five more pages to go.

I've got to get on so that I can get on to book six.

I've skipped lots of books y'know.

Most of the children were very keen to race on through the schemes' workbooks. It was not that the teachers urged this rush. On the contrary they persistently emphasized steady work and had to remind the children they were to do only as many exercises as they had been told. When the brake was applied it worked for a little while. Robert, for example, was told — in common with his group — to do the first two sums in an exercise. When asked by the interviewer why he had not done the whole sheet he looked horrified and said, 'I'm not allowed to do that!'. And even Gemma, who was ever willing to be creative, was seen to do eight sums in an exercise and rub four out before going to the teacher. She had evidently recalled — in the nick of time — that Mrs. F. had ordered only four sums to be done.

Teachers' reminders were soon forgotten unless persistently repeated. Once forgotten the children got on with what appeared to be their main business, catching up with or better yet getting ahead of their friends. All the children knew exactly where their friends and rivals were in the schemes and most were anxious,if not to get ahead, at least to stay in contention with what they seemed to see as a competition. This appeared to be a self-imposed race because we saw not a single instance of a teacher ever making any comparisons between children or ever extolling the virtues of speed.

Whilst mathematics educators worry about the preparation of

children for the adult world, such a world had no reality to these youngsters. They never saw adults doing mathematics and in any event they were far more concerned with the reality of their own world which, mathematically speaking, was the self-imposed business of getting on through the scheme. There were no ultimate prizes for this — or at least certainly none from the teachers. The race was the thing and providing you pushed on you were winning. In this light it is easier to understand the children's unwillingness to rate discussion sessions and their lesser interest in them. Perhaps the children saw them as a distraction from the main agenda.

Only one child rejected the validity of the mathematics scheme. Six-year-old Robert suddenly became adamant in the belief that it was all a waste of time:

> . . . there's more stuff than that. I burnt my finger yesterday and you should learn about that and you should learn about cookery.

Overview

The children toiled eagerly on their paper work. They appeared to be embroiled in a race to get through the scheme as fast as possible or at least to stay in contention with their chosen rivals. The competition was not endorsed by the teachers who persistently applied the brakes by demanding limited production, checking, attention to neatness and so on. In terms of work completion there was a high level of success based in the main on familiarity with the task demands. The children did not indicate this familiarity to their teachers. This avoided the risk of making their lives unnecessarily difficult. The teachers did not find out about the degree of match between the tasks and children's attainments because they did not conduct any detailed diagnostic work. If a child could do a task all was a well; if not he was re-taught.

The children were much less interested in discussions. There was a greater level of fidgetting and the children never spontaneously referred to these sessions.

The content of the work was dominated by the commercial schemes used, in part because the structure of the schemes looked like a race track to the children and in part because, forming the only overtly assessed part of the mathematics curriculum, they could be taken to be the whole of what was meant by mathematics. The children did not see the relevance of this work either to the adult world or to their own world outside school. The schemes could be seen as a significant mediator of the children's mathematics experience.

Whatever the shortcomings, there seemed every reason to believe that children's pleasure in basic number skills was being fostered. This was presumably very reinforcing for the teachers' approach.

Chapter 6

The Teachers' Perspective

It has already been shown — in chapter 3 — that the massive logistical problems of the early years' curriculum had received from the teachers a great deal of attention followed by intense industry as they prepared the management of their classrooms. It will be shown that the problems of delivering the curriculum, in flight as it were, received no less a response.

Data on the teachers' thinking, planning, reactions, evaluations and in-process decisions were gathered from a range of sources. Conversations were held at a number of points in the research programme. These included pre-session discussions about aims, methods and prospective problems, post-session discussions about performance, specific issues, particular children and their work and lesson evaluations, and ad hoc conversations about crises, or emergent points of interest. In addition to these sources of data it will be recalled that thirty-five mathematics sessions were recorded on video tape. Each of these tapes was the subject of extensive discussion with the teacher concerned.

Analyses of these conversations were used to build a picture of mathematics teaching from the teachers' point of view. What emerged was an image of bewildering complexity and exquisite sensitivity as the teachers responded to what they saw as a bombardment of information and an incessant demand for decisions.

Before entering the classroom, all the teachers reported that they had made a number of decisions which, were they to be followed through, would fundamentally affect the conduct of their classes. These decisions referred to aims, seating arrangements, content, use of materials and their children's previous knowledge. Whilst not all of these areas were consciously debated prior to every session, each had, at some time, been exhaustively considered.

All the teachers reported making decisions about their aims for each session. Sometimes these were expressed as simply procedural aims 'I aim to introduce them to the idea of a simple addition sum', others were driving at concept acquisition — 'I hope to establish the sixness of six' whilst more frequently the aims were a complex mixture of these notions, 'I want them to get an early understanding of place value and what tens and units mean. I also want some of them to think about recording it.' However briefly expressed, it was clear that the aims had all been clearly thought out and were in fact only a guiding outline for a lesson rather than a fixed end point. As Mrs. E. put it, 'I have different aims for different children.' — And Mrs. F. noted,

> I want to cover the simple addition of two digit numbers. I'll have to check that they can remember how many tens and how many units there are in a certain number and I'm not going to think beyond that first step because if I desperately want to get them on to adding tens and units it would just be my luck that they won't remember the basics — how many tens and units in thirty-six.

The aims were thus seen as merely a general plan which would have to be particularized for specific children and almost certainly amended in the light of the action in a session.

All the teachers gave detailed, pre-task consideration to the form of grouping they would use for the children. These decisions of general management have been discussed in chapter 3. On a day-to-day basis however, the allocation of children to groups presented the teachers with ever changing problems. For example, Mrs. D. had great difficulty in placing Caroline — a very vocal seven year old — who frequently 'says anything that comes into her head'. As Mrs. D. put it, 'She is currently in the blue (middle) group but I've had her in the red (top) group and I've had her in the green (bottom) group and now I'm wondering whether to put her in the green group again. It is just so difficult to know what she knows. Is she just impulsive or is she really not very able?' In Mrs. B.'s class Joanne posed a different problem, 'Strictly on ability she should be in the bottom group but I prefer her in the middle group. She fits in so nicely with that lot and it's a social thing as well as an intellectual thing that I want to get going.' Mrs. E. saw Danielle as an able but very quiet child and considered that she would have to be placed in a group which would allow her to express herself — and especially her mathematical ideas. After much thought and some trial and error Mrs. E. decided to put

Danielle with just one other child — Toni — whom she felt would be 'an intellectual match and a suitably responsive and sensitive partner'. It was clear that whilst, as a general principle, the teachers who grouped children did so on the basis of general ability, it was recognized that this was not always easy to determine nor was it the only factor considered. Children's personalities and ways of socializing (and frequently their ways of not socializing) were sensitively considered and attempts made to use peer group processes to help children make the best of themselves.

This sensitivity to children's needs was extended to the consideration of the use of appropriate materials in the lessons. The teachers' broad principles of logistics were discussed in chapter three. Again, however, general principles were complexly interpreted in the light of individual cases. For example, teachers of third year infants frequently felt that whilst some of their pupils still required experience with concrete apparatus there was a strong tendency for the children to dismiss this as babyish. Mrs. D., for instance, noting that Richard was struggling with number bonds to six, thought that he needed, 'something that's really special and very different. There's no good my going to get the animal shapes out — they used them ever so much last year for counting — because Richard would just dismiss that.' This was only one of a multitude of instances in which teachers commented on their children's differential responses to apparatus according to their personalities and this was an important factor which they considered in their assembly of materials.

In response to their teaching problems it would seem that whilst the teachers never saw themselves as quite winning they felt that they were certainly not losing and in their ever thoughtful responsiveness to children they valued their capacity for quick recovery at least as much as their capacity for forward planning. The former skill was perceived to be crucial in response to any change in routine. For example, whilst seating arrangements for the children were very predictable (they sat either at their desks or in a group on a carpetted area) it was sometimes felt necessary to alter this organization for the purposes of a particular demonstration. Mrs. G., for instance, wanted to discuss three-dimensional shapes with her 6-year-olds and she was anxious to capitalize on the practical work the class had done on the previous day. In planning the lesson, she decided that — instead of sitting on the carpet as usual — the children would sit on chairs in a semi-circle round a wall display of their own work. In the event the children, unused to sitting quietly on chairs for any length of time, became very disruptive; so much so that on seeing the video recording of the session, Mrs. G. observed, 'I am surprised I have not killed Ben'. Whilst, as she noted, she would think carefully before changing the children's routine again she also pointed out that had she not done so in this instance she would have had to have forfeited the use of the established wall displays. This was only one instance of the endless stream of dilemmas the teachers met each day. Every solution created its own new problems.

Complex though these preliminary planning decisions on aims, materials and groupings are, they can be taken before any action starts and in that sense they can be informed by a relatively leisurely reflection on aspirations and past experience. Once the children arrive such leisure becomes impossible. Information must be processed much more rapidly.

A class of young children invariably arrives with a set of cuts, bruises, upsets, elations and other distractions. The teachers considered it essential to make a rapid assessment of the class's mood and receptivity. When Mrs. E. introduced some new furniture into the play area the children 'got very high' in anticipation of using it. She therefore decided to postpone her mathematics session in order to capitalize on this interest. Simon entered Mrs. C.'s class in tears. She felt he would not be able to focus on his work until he had unburdened himself so she set her plan aside and spent several minutes listening to his troubles whilst the rest of the class chatted amongst themselves. Such major distractions were not common. In the main the teachers had long established routines for moving into work. However the teachers always reported monitoring the mood of the class. Their comments revealed that whilst decision times were drastically reduced compared to those involved in pre-lesson planning, they lost none of their responsiveness, humanity and gentleness with children. Nor did they lose sight of their aims. Whilst prepared to tack (and occasionally heave to) in response to their children, the mathematical aims were never set aside for long before the skills of diverting children's attention on to their work were brought into play. Indeed it should be emphasized that the teachers' major reported preoccupation was with delivering the mathematics curriculum. The diversions have been reported for what they reveal about the teachers' information processing. The information processing was typical. The diversions were less so.

Once the mathematics action started there was a quantum leap in the reported intensity of the information processing in the form of problems to be solved and dilemmas to be resolved. For example, Mrs. G. was discussing the demonstration sheet shown in figure 13.

Figure 13: Place value demonstration sheet

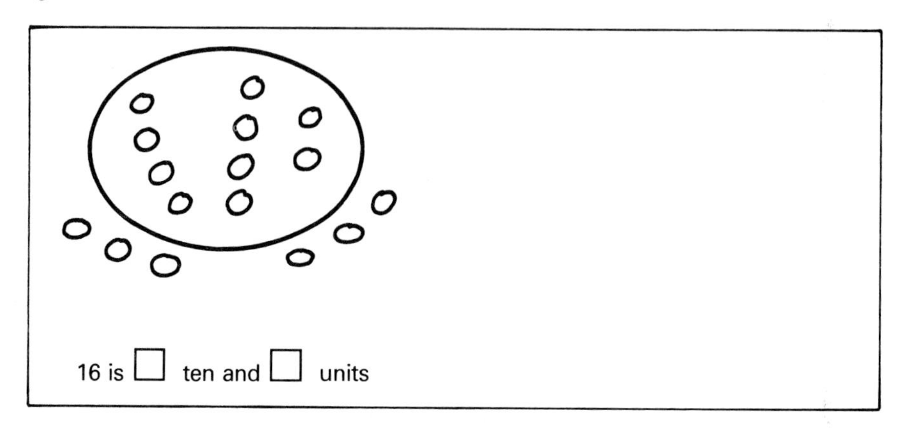

She turned her attention to Michelle who had her hand up.

Mrs. G.: Michelle, what number should I put in the first box?
(Michelle looks horrified.)
Mrs. G.: Ah panic! Don't you want to do it? Come and try.

At this point Mrs. G. recalled thinking,

> She panicked when I asked her. She is very keen to do things but she did not want to be wrong. When I saw her face I thought, 'Now shall I suggest she doesn't do it?' But then I thought it would be a good moment to try and move her on a bit. So I decided I wanted her to see it through. Because even if she wasn't very sure of it it would be a good teaching point for her and she would learn better by that rather than somebody else helping her do it.

The other teachers frequently expressed concern about their pupils' confidence and emphasized the problems involved in trying to strike a balance between developing children's independence and maintaining their confidence. Mrs. F., for example, noted how she carefully gauged how long each child might be given to answer a question, 'You can't say, "after two and a half minutes I will interfere." In two and a half minutes, Elizabeth would be hysterical whereas someone else might say, "I'm still thinking." It depends on the individual child.'

In one particular case in which 7-year-old Roscoe appeared unsure about how to tackle a task, Mrs. F. commented,

> He's very unsure. He's looking for reassurance. I try to keep a blank face. It's very hard and it's very hard for the child as well because there they are, staring up at this po-faced person who is not giving anything away. It's very difficult because your instinct is, particularly with a child like Roscoe who hasn't got a lot of confidence, is to interfere and to help. But interfering is not going to help him. You can only help him by solving his problem and the way to do that is to let him work it through. It's difficult. It's ever so difficult.

Mrs. B. was faced with two different dilemmas early on in her lesson on addition. After explaining the procedure for three minutes she was aware that two members of the group still did not know what to do. Instead of prolonging the oral part of the lesson, she decided that the children should set to work individually.

> I did not want to keep the rest of the group waiting and although I know that two of them did not fully understand, I thought it better to go on to the written work rather than risk boring them to tears.

A few minutes later, Mrs. B. was aware that both Thomas and Andrew wanted her help,

> I was sensing that Thomas was coming up to a crucial point where he was going to understand. I knew that Andrew was also so — 'oh dear, what should I do?' But I think Thomas was in more urgent need of help at that particular second in time. So I decided that Thomas had to be dealt with and Andrew would just have to wait.

These incidents are taken from a stream of events and reflections. They illustrate the concerns and considerations of the teachers but they do not remotely convey the teachers' sense of being bombarded from all sides with information, distractions, and interruptions and of feeling under relentless pressure to make decisions and resolve problems. The following paragraphs give some image of one teacher's perceptions of this challenge. The extract below represents one minute of a session in which the teacher had been following a routine procedure to get children to demonstrate 'number stories for ten' that is to break ten down into two component parts.

> *Mrs. D.*: Make a ten stick of cubes that are all the same colour. (Children do so.) Right. Check that you've got ten. Now we are going to make number stories that add up to

ten. Now show me with your cubes a number story that adds up to ten.
(The children break up their sticks and mutters of 'one add nine' and 'two add eight' can be heard. Ross breaks his into two fives, looks round, re-builds his ten and breaks it into one and nine.)
Mrs. D.: No Ross. You were right. I just said, 'show me a number story'.
(Ross reproduces two sticks of five.)
Mrs. D.: Now. What have you got Ross?
Ross: Five add five.
Mrs. D.: Five add five is ...?
Ross: Ten.
Mrs. D.: Right. Ten. Hazel what have you got?
Hazel: One add nine.
Mrs. D.: One add nine makes ...?
Hazel: Ten.
Mrs. D.: Ten. Right. What have you got Kai?
Kai: Three add six equals ten.
(Kai is holding up a stick of six and a stick of four.)
Mrs. D.: Three add six equals ten. Noooo ... Have another look.
Kai: Four add six.
Mrs. D.: Four add six makes ten.
Robert: Two add seven.
Mrs. D.: Two add seven makes ten? You count again. What do you think they are Darren S.?
Darren S.: Two add eight.
Mrs. D.: Two add eight. What have you got Darren L.?
Darren L.: Three add eight.
Mrs. D.: (Waving away Freddie from another group.) No. Not three add eight. What have you got Richard?
Richard: One add nine.

Before examining the teacher's comments on her decisions during this extract it is worth noting that she had thirty-five children in a very small room. She was working with one of three groups. Having organized their work she had to decide which children to call on. Her decisions were influenced in part by her knowledge of the children which had been acquired over the year and in part by the momentary changes in the events which unfolded before her. Mrs. D.'s comments on the action were as follows:

> I saw Ross do something unusual in making five add five. As soon as someone said 'one add nine', he (Ross) started to put his back. I was also trying to see what everyone else had done and who was looking at whose. Louise, Hazel and Lindsay are never too sure. I saw Hazel had it right so I called on her to give her a bit of confidence. I saw Kai had it right. I am having terrible battles with him because he won't write things down — he won't finish, especially if he thinks you are watching him. I was not surprised when he said it wrong even though he had it right. Robert was the same. He had three and seven and said 'two add seven'. He and Kai both knew the larger number but probably guessed the smaller number instead of counting it. They might expect to know two and three and four without counting. The work is totally inappropriate for Darren L. He should not be in this group but I have to keep an eye on him. There was no point in going on about his error at that stage. I waved Freddie away without a second thought — he knows the rules. Richard is a bit like Darren L. I am not sure he should be in this group but I saw he had it right and that he was not paying attention, so I got him to announce. He is a puzzle. He knows loads of number bonds but cannot count consistently up to ten. He says he likes adding up but does not like counting. He is difficult to talk to because I cannot follow what he says.

In this one brief instant of teaching a routine task the teacher found herself sustaining the general pace and direction of her lesson, pushing for completion, monitoring individual responses and comprehension, choosing one child to boost confidence another to check understanding and yet another to attract attention. All the while she found it necessary to interpret their particular responses in the light of her knowledge of their typical behaviours. The wrong answer of one child was judged worth correcting immediately whilst that of another was judged better left to another occasion. These complex interactions are shown in diagrammatic form in figure 14.

Whilst the diagram is complex — and doubtless less complex than the mental processes it is intended to represent — Mrs. D. found all this relatively normal. Her account is typical of those given by teachers as they conduct what they consider to be routine lessons. The demand for decisions is perceived to be incessant — and exhausting. It is small wonder that each of the teachers reported that they occasionally went into 'automatic pilot' and held a period of routine questioning

Figure 14: Some of the factors (and the interrelationships between them) Mrs. D. considered during a minute of her teaching

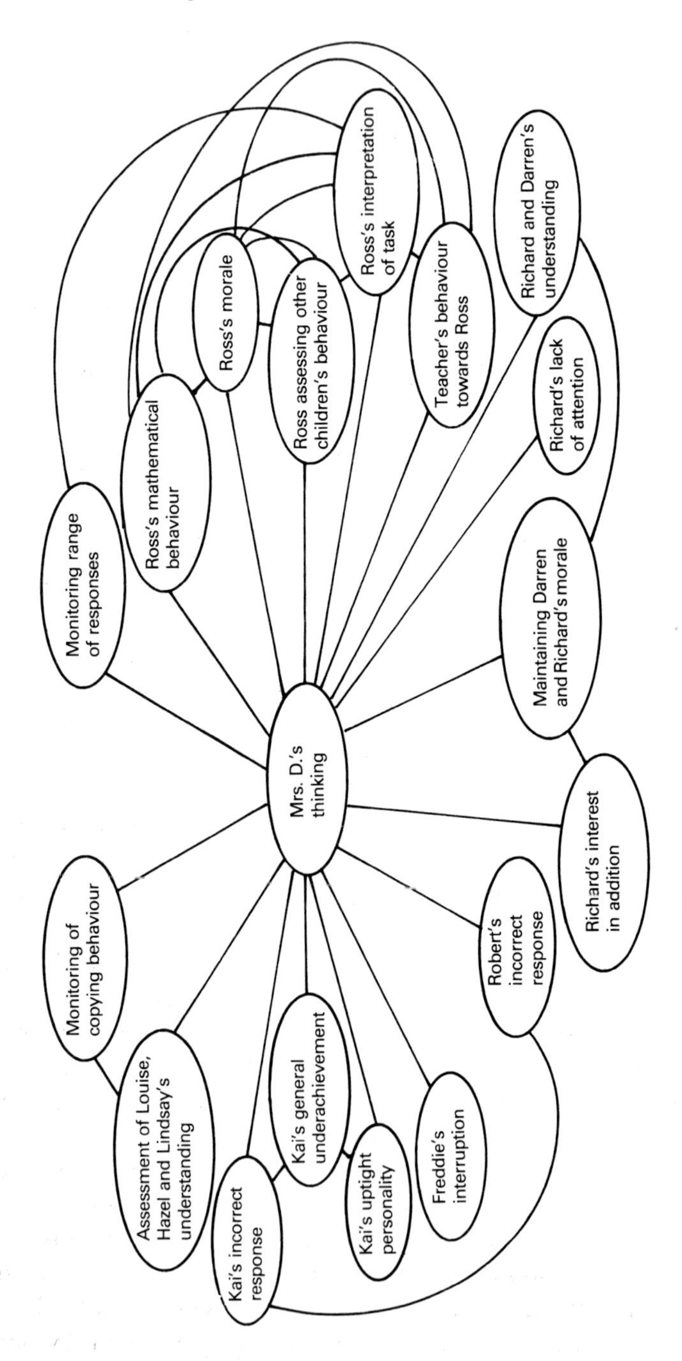

in order to, 'have a bit of a rest for a second' or, 'to give my brain a breather'.

Granting the speed and complexity of the issues the teachers nonetheless reported relatively few instances in routine sessions when they had, in their terms, to make 'major decisions'. As soon as they stepped into more exploratory work however, they reported facing yet more challenging problems. This is illustrated in the following account. Mrs. G. set out to introduce her third year infants to the names of three dimensional shapes. She had the children sitting in a circle on the carpet. They had established that a globe was sphere shaped. A globe had been passed round. Its spherical properties had been pointed out and felt.

> *Mrs. G.*: Sphere. That's right. Now can anyone else tell me anything that's a sphere? Adam? (who has his hand up)
> *Adam*: Square.

> At this point Mrs. G. recalled, 'I was ever so surprised he said that. I nearly fell off the chair. He is so keen and so shy. I thought, 'how on earth am I going to cope with this without putting him down, without smashing the bit of confidence he has got? It really was about the most inappropriate thing he could have said.'

She decided to ignore his answer and to get Adam to feel the globe and describe it as a sphere. She then turned her attention to another child. Seconds later, in giving a further example of an everyday object that is a sphere, Ben offered 'Half the world.' Mrs. G. noted

> I thought, oh dear! I cannot get into all that now. If he had said it later when we had got the main idea established I could have developed it using plasticene. As it was, lots of the children were just beginning to get the idea and I thought it would be too confusing. I suspected he (Ben) had the idea and would have benefitted from some extension but I decided to pass over his suggestion for the sake of the rest of them.

There followed a ten second period of acceptable responses for sphere shaped objects and then Samantha suggested, 'a circle'. Mrs. G. thought,

> 'Help! This is going to be a lot more difficult than I thought. How am I going to explain the difference between a circle and a sphere?

In an attempt to do this she asked Samantha to put her hands all around the globe and then asked her to compare that with a circle (a hula hoop on the floor). As the child did so it occurred to Mrs. G. that Samantha would be able to feel the depth of the hoop and she thought,

> We were getting deeper and deeper into it. What could I use? I haven't got the time to cut a circle out of paper. I'll have to make do with the hula hoop and hope she sees the difference. If I cut out a circle the point might be lost and the group might lose interest.

Throughout the next minute the children gave examples of spheres, all of which involved the idea of a ball; a cricket ball, a golf ball, a football and a bouncing ball were all offered. It looked as if things were back on an even keel when Gary suggested, 'a head'. Mrs. G. experienced her fourth dilemma in under two minutes.

> At first I was pleased. That was a nice idea. It was quick to relate it to spheres. Then I thought that a head is not a perfect sphere. It was really difficult to balance up because I did not want them to get wrong ideas at the beginning — they are really capable of that. At the same time I did not want to bog them down with pernickety bits and pieces. It was hard to sort of weigh that out. I was really struggling.

In this brief span, Mrs. G. had perceived that she had to resolve a number of difficult problems involving conflicts of concern for mathematical concepts, children's personalities and interests, individual and group needs, the management of materials and for timing. Once the teacher engages in relatively free discussion with the children, pupils' responses become unpredictable and the teachers perceive themselves as having to make the best of what they get. Levels of uncertainty, ambiguity and risk are perceived to rise drastically and so much so that a few moments in 'automatic pilot' become increasingly attractive.

Problems in Information Processing Under Time Stress

It has been shown that the teachers set out with clear aims but an open mind as to how these might be achieved. Their conceptions of ends and means were derived from a relatively leisurely and detailed consideration of the demands of the mathematics schemes, available apparatus, the management of children and materials and of the children's attainments in mathematics. Teachers' comments on this stage of their thinking indicate extensive knowledge of their material and apparatus and how children interact with it. They also reveal a fine awareness of their children — not merely as mathematicians but as humans with sensitivities and worries as well as academic strengths and weaknesses. At this point the teachers' detailed professional knowledge is well matched by their behaviour as evinced by their copious preparation.

Once the teaching action starts it has been shown that the quantity of information the teachers have to handle and the rate at which it must be processed increase dramatically. Particular moods, events and especially children's responses — each in themselves potentially difficult to deal with — come in a welter of complex interactions. In routine lessons teachers consider all this as fast and furious but relatively normal and containable. Experience and good management puts them in a position, as they see it, to react profitably. In open discussion the predictability of children's responses is low and the probability of having the appropriate material to hand to capitalize on the children's contribution is even lower. In consequence dilemmas proliferate in the teachers' minds. Their comments on their thinking during these interactions reveal doubts and hesitations. What they do not reveal is an awareness of the increasing mismatch between their thinking and their behaviour.

At all levels of activity the teachers' essential humanity to children was manifest. Unfortunately, at levels of intense information processing, another aspect of humanity begins to take its toll. It has been well established that in novel situations humans cannot successfully process very much information at once. If problems come in quick succession and decision times are short, human reasoning tends to exhibit certain well known and — in pedagogic terms — unfortunate characteristics. Laboratory studies of problem solving under stress and real world studies of airline pilots, surgeons, policemen, fireman, businessmen and others in emergencies show that thinking exhibits some or all of the following characteristics. There is a narrowing of the range of possible responses considered; there is a failure to notice certain events, or if noticed, a failure to realize or act on their significance; there is a persistence and repetition of some behaviours in the face of failure and there is a reluctance to test hypotheses or hunches. Ideas which might normally be seen as speculative are seen, under time constraint, as certainties and evidence which contradicts these 'certainties' is simply denied. These pathologies of reasoning have direct, and sometimes dire, consequences for behaviour. Apparently 80 per cent of airline crashes are due to pilot error — frequently in cases where the pilot has had plenty of warning from his instruments that all was not well. It appears that interpretations of these warnings, under stress, exacerbate rather than alleviate the predicament.

In the much more mundane circumstances of laboratory problem solving it has been reported that,

> In some of our research we were forcibly reminded of pathological phenomena which are not normally found or looked for in psychological experiments. Repetition, asseveration, self contradiction, outright denial of the facts and ritualistic behaviour became quite typical in some tasks ... these symptoms were reflections of reasoning under mildly stressful conditions ... (Wason and Johnson-Laird, 1972)

Lest it be thought that anything dastardly was going on in these experiments it is worth noting that the tasks were relatively straightforward. In one task, for example, undergraduates were told that the three numbers, two, four, six conformed to a simple relational rule and that they were to discover this rule by generating successive triads of numbers. Each time they were told whether their triad conformed to the rule. The somewhat obsessive behaviours described above were frequently seen in response to this task. The researchers were not implying that their subjects were pathological — only that under these apparently simple circumstances their behaviour had some resemblance to pathological thought. The work of Wason and Johnson-Laird is consistent with a great deal of research on problem solving which shows that if people are left unaided with a problem, their thinking frequently shows 'fixity', that is to say, a tendency to persist with unsuccessful approaches and a denial of or blindness to alternatives. Their behaviour may show all the characteristics of stupidity or ignorance. Time pressure is known to exacerbate all these effects.

It is not surprising that all these features were evident in the behaviour of the teachers studied. The less routine and familiar were the children's responses, the more the teachers' behaviour became fixated. The crucial thing to emphasize is that there is nothing remotely unusual in this. It simply illustrates that the teachers are human.

The teachers frequently rushed to interpret children's responses and did not pause to check or test their interpretations. In one instance, Mrs. F. was explaining how to complete sums of the following type:

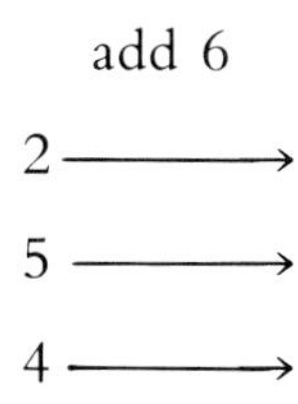

She noticed that Vicky was writing '6' at the end of every arrow and concluded, 'She is clueless, absolutely clueless. There is no understanding at all there.' A post-task interview showed that Vicky had a good grasp of addition but did not grasp what the card required.

The rush to interpret correct answers was even more evident and more frequently in error. Time was often spent practising a procedure until the teacher was satisfied that the children understood the underlying idea. In her attempt to get children to understand subtraction Mrs. D. had them read out subtraction 'number stories' from diagrams like those shown in figure 15. (The details of this event are in chapter 4.)

Figure 15: Sets for subtraction number stories

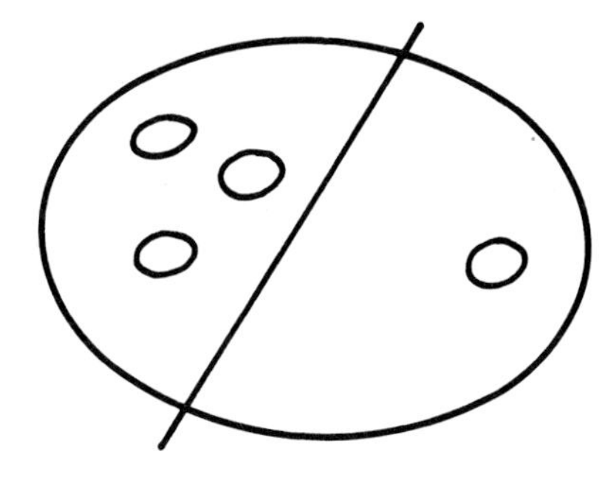

After some initial hiccoughs all the children were able to say, 'four take away one leaves three' in response to this diagram. Mrs. D. concluded that they, 'had got a good grasp at last'. In fact post-task interviews showed that just fifteen minutes later and without Mrs. D.'s conducting they could not reproduce their behaviour.

Also seen was the repetition of unproductive teaching behaviour in response to a child's failure. In the following example Mrs. B. was trying to get Michael to read out the number symbols to five:

Mrs. B.: Now Michael. What number is this (points to '5').
Michael: Number two.
Mrs. B.: Number?
Michael: Number . . .
Mrs. B.: You count. (She points to the numbers as Michael says them)
Michael: One, two, three, four, five.
Mrs. B.: What number is that? (pointing to '5')
Michael: Number four.
Mrs. B.: You count again (points)
Michael: One, two, three, four, five.
Mrs. B.: What number is that? (pointing to '5')
Michael: Number four.
Mrs. B.: Start again.
[The cycle continued three more times until Michael said 'two' as Mrs. B. pointed to '5']
Mrs. B.: Well what number is that? (pointing to '2')
Michael: Number two.
Mrs. B.: Well does that number ('2') look like that number ('5')?
Michael: [no response].

Obviously Mrs. B. was fixed on a routine that was going nowhere. One repetition of the cycle would be understandable; five repetitions seem distinctly unprofitable. Why did the teacher so persist? Her own account of her thinking during these events shows that she was surprised, even dismayed, by Michael's responses. She felt, '*surely* he would see . . .' that having said 'five' when pointed to '5' he would say it again when she pointed again. And 'surely' after two or three repetitions this would 'click'. Mrs. B.'s behaviour was sustained by a very powerful view of Michael's problem as a learner. He was seen to be inattentive. This was not considered a hunch to be tested. It was treated as a certainty. The problem, in Mrs. B.'s eyes, was to focus Michael's attention on the salient cues. A lot of the moment-by-moment evidence available to Mrs. B. was entirely consistent with this view. The video tape shows Michael's eyes did wander. He fidgetted a lot. A non-participant observer might easily have interpreted that as signs of stress or discomfort. But that amounts to the same thing — inattention.

Eventually Mrs. B., sustaining the same hypothesis of inattention, decided on a different tack and took Michael to a set of number symbols cut out in sandpaper and had him trace round them with his finger observing that, 'If I cannot get to him through his eyes and ears I will have to go through his fingers.'

One hypothesis that Mrs. B. did not entertain was the possibility that Michael had a very shaky grasp of quantity in this range. Even had she got him to say 'five', on this hypothesis it would have been a largely meaningless achievement. In fact the pre-session interview suggested some such interpretation of Michael's competence. He could recite numbers to five but he could not count off more than three objects if they were of the same type. He could not select three objects from a group. He could not match symbols to quantities in this range. What Mrs. B. was asking Michael to do was not only to count to five (which he could) but also to realize (because it was never pointed out) that the symbols related to the words said. This he could not do. Had Mrs. B. known more about Michael's competence before this task she might not have expected him to attempt it. Indeed, without a lot more concrete experience of quantity, success for Michael on this task at this stage would have been merely barking at print.

It would appear that the teacher did not know about Michael's strengths and weaknesses in regard to this task and was unlikely to find out given her hypothesis about his failure to attend. Given also that there was plenty of evidence consistent with her hypothesis — evidence seen as confirmatory — alternative explanations were unlikely to be conceived, let alone tested. The force of Mrs. B.'s hypothesis was such that she did not even consider the strategy of leaving Michael alone in this instance. His attention had to be locked-on one way or another. She was dismayed that a tried and tested strategy of repetition had not worked. This was unusual in her experience. Whatever it does for children's understanding, repetition usually gets them to say the right answer. Once she had got over her dismay at the failure of her first tactic she showed some inventiveness in adopting another (albeit within the same strategy). However, Michael's basic problem remained untouched and increasingly untouchable as his teacher narrowed her aim to that of getting him to say the right thing at the right time.

What is noteworthy about this instance is that it was commonly seen (in response to children's errors) in all its main features including the narrowing of the teacher's aim, the repetition of a failing tactic and the absence of an attempt to step back and take a longer look at the problem or to test the basic hypothesis on which the teacher's behaviour was based. In short the teachers, in response to novel behaviour by the children evinced all the characteristics of dull and uninformed people. The more open the teaching, the more novel was the children's behaviour and — in train — the more problems were perceived and disappointments experienced.

That these teachers were certainly not dull has been demonstrated repeatedly in these pages. When given the opportunity to comment on the recordings of each others' sessions they showed all the flexibility and insight necessary to generate and test alternate views on children's contributions. Their limited thinking in action must be attributed in part to the press of events. To adopt fresh lines of approach several notions must be grasped rapidly if continuity, attention and interest are not to be lost. The necessity for a change must be perceived. A new angle must come to mind and a justification must come with it. Otherwise there would be little point in the change. This is a very tall order indeed particularly if the reader recalls what the teacher can never forget — the teacher and child are never alone in the classroom.

The press of events however, cannot be the sole reason for the teachers' responses in these circumstances. From the whole range of events happening at any moment, the teacher makes her selections and we may presume that these selections are based on preoccupations that the teacher brings with her to the situation. These preoccupations were not identified as such by the teachers we studied. They did not emerge spontaneously nor were they forthcoming in response to our questions. We are therefore left to speculate as to what ideas, under circumstances of time stress, lure the teachers away from their principles and into fixations.

Once such notion seems to be that once an activity has started it must be completed. In the above example, Mrs. B. seemed unable to step back and leave Michael to give herself time to think.

This unwillingness to step back might in turn be bolstered by the notion that children are extremely vulnerable both in terms of their confidence and in their capacity to be led astray by error or wrong ideas. Teachers' references to the need to sustain children's confidence permeated our discussions. Of course at the level of general principle this is unimpeachable. But at the level of practice, as Mrs. F. pointed out, to sustain confidence it is extremely difficult not to jump in and help at the first sign of a child's hesitant glance. As the teachers recognized, the price of confidence can be dependence but it is a price often paid in the maelstrom of classroom exchanges in which teachers give in to the human instinct to help in preference to their own pedagogic strictures. Confidence in this view demands a good result quickly.

Confidence is not the only issue at stake. It would seem that the teachers had the idea that children can be easily misled into false notions and inappropriate routines. Mrs. G.'s unwillingness to explore her children's examples of sphere is an instance of this principle in operation. She recognized the fertility of some of the children's ideas but '. . . passed over discussion for the sake of the rest of (the children)'. If a teacher holds the view that wrong or unusual ideas can be very misleading then she must endeavour not to start them (*q.v.* Mrs. G.) or, if started, to sort them out as soon as possible (*q.v.* Mrs. B.). Both responses are premised, logically, on the view that there are 'wrong ideas' or that there are dangerous routes to 'right ideas'. Now clearly, these teachers did not hold either of these views. In chapter 3 we showed how highly they regarded children's ideas and the potential for sharing and exploring them. They doubted only their own capacity to manage children's intellects as a resource in the classroom. It seems that children's ideas are indeed very difficult to manage and they rarely got on the agenda in the classroom because the agenda is already crowded by something that is easier to manage, namely the mathematics scheme. And in fact the schemes do imply that there are right ways to operate and proper uses of terms to be shared. It seemed to us that once the teacher is committed to using the schemes as published — and there are many attractions to this option — she might well feel committed to the behaviours we reported in chapter 4 and the mental life we have reported here.

Summary

Teachers reported that they were highly sensitive to a wide range of interacting factors when in the process of teaching mathematics. Amongst these factors were, their aims for the session, the children's personalities (especially their confidence), group interactions, materials used, seating arrangments and the particular responses of children. All exchanges demanded persistent monitoring and posed a string of problems to which the teacher perceived no ideal solutions. The more the content of the sessions moved towards new or unfamiliar work the more acute the information processing demands became on the teachers. Under these circumstances the teachers became more hesitant and were seen to act in ways which ran counter to their principles and of which they seemed hardly aware. Their behaviour became consistent with the view that they were operating on the unarticulated principle that the assigned work must be completed in the manner indicated by the schemes. If this view is accepted we see once again the powerful role played by the schemes in the determination of the quality of mathematical learning experiences.

It was clear that whatever the content of the exchanges, wherever teachers attended to more than one or two children whilst at the same time as monitoring the rest of their class, they had to contend with a vast amount of information and make very rapid decisions about their responses. These constraints seemed to force the teachers to be less rather than more open ended in their thinking. In short, the teachers behaved like humans.

Chapter 7

Understanding the Teacher

In this chapter it is our intention to draw together the threads of chapters 2–6 to fill out our account, in so far as our data and intuitions permit, of the teachers' behaviour in their classrooms and in particular of their relative lack of attention to inculcating the higher order skills of problem solving, applications work and open-ended exploration.

The achievements of the teachers were considerable. Their children were almost invariably happy in the classes. They were seen to get on with their mathematics work and were confident at it. They appeared to enjoy their mathematics and, particularly with respect to the paper work, were generally anxious to do more. Whilst we were not in any one classroom long enough to measure any learning gains the teachers and children were observed indulging in those work habits which previous research has shown to be associated with good progress in learning. The teachers' actions were highly consistent with the model of direct instruction described in chapter 1 and the children spent a great deal of their allocated mathematics time concentrating on assigned tasks and achieving a high success rate. We might anticipate that these children would have made progress in their mathematics at least in so far as progress is conventionally assessed.

Whilst working on their mathematics the children had learned a lot more besides. They had learned how to organize their work, how to obtain, use, share and return equipment, how to use classroom resources and how to relate to the teacher and other children, as occasion demanded, to sustain their activities.

The limitations in these classrooms were equally evident. Instances of every matter of concern expressed in the Cockcroft Report could be readily seen in each case. The activities were dominated by pencil and paper work in which the children practised routines explained by the teachers. There was very little investigation or applications work, real problem solving or pupil-pupil discussion. Children's spontaneous approaches to calculations or procedures were overlain with teacher-directed routines. Except when calculations were made to refer to sweets and such-like the content of the curriculum had little relevance to the children's world out of school — a fact clearly evident to the children themselves (although taken by them to be a matter of no importance). There was very little evidence of any technology applied to mathematics. We saw no hand-held calculators in use for example. About one-third of the work assigned seemed beyond the children's comprehension although in about half of these cases the children could complete the exercises by some means or other. The children's commitment seemed to be sustained by large amounts of praise for effort or for the completion of exercises. The teachers rarely indulged in diagnostic activities. They preferred to meet mistakes with repeated instruction.

The behaviour in these classrooms was very similar to the behaviour reported in scores of other studies on the teaching of mathematics and almost identical to that which we saw in our previous study of infant classrooms (Bennett *et al*, 1984).

Despite the considerable achievements in terms of the acquisition and exercise of basic skills and facts there was a shortfall on aspirations as expressed by mathematics education experts. There was apparently little attempt to develop children's intellectual autonomy, to develop strategic skills related to problem solving, to develop skills necessary to share and evaluate ideas or to consider how real problems might be approached.

These deficits could not be due to any lack of qualities on the part of the teachers. We showed in chapters 2 and 6 that the teachers were perfectly aware of the aspirations held for mathematics education and that they fully endorsed these aims. We showed in chapters 2, 3 and 6 what massive personal resources they brought to bear on their mathematics teaching. The teachers had enormous respect for the quality of children's thinking and the extent of their knowledge. They showed considerable pleasure whenever they saw these qualities exhibited. Their respect was not based on the rather limited findings of developmental psychologists but on their own more extensive experience with children.

The teachers had an elaborate view of learning which saw a place for enquiry, practice and problem solving and they articulated the range of teaching behaviours necessary to make the most of such a multifaceted theory of the acquisition of knowledge and skill.

The teachers further demonstrated the qualities of their own insight during the analysis of video tapes and in their comments on the mathematics schemes. Their capacity to spot problems, generate hypotheses and weigh alternative considerations left little to be desired. They were acutely aware of the social and emotional dimensions of their children's personalities and how their needs might be met in terms of the organization of groups and tasks in the classroom. The teachers cared intensely about their children. This care was expressed not in a sentimental way but in a thoroughly professional response.

The teachers were more aware than any researcher of the variety amongst the children in their charge. Indeed, the psychologist's image of intellectual diversity was elaborated to take into account the interacting effects of personality and social behaviour. The challenge of the range of mathematical attainment as decribed in the Cockcroft Report was as nothing compared to the challenge of the interactions between attainment, emotional maturity and general socialisation. Where researchers report 'dimensions' of differences these teachers saw kaleidoscopic patterns.

The teachers were no slouches at preparation. The mathematics work was only one part of the material necessary to keep thirty children busy throughout the day. The arrangement of supplies, the provision for transitions and continuity, the planning for distribution, action, collection, display and the demands of record keeping in all areas of the curriculum had to be met. We saw not a single hiccup in these matters during our period of observation.

These were clearly teachers of considerable quality. Still they failed to deliver on important aspects of the aspirations held by mathematics educators and which they themselves fully recognized and endorsed. How did their relative failure come about? Why was there so much pencil and paper work and so little meaningful investigation? Why was there so little teacher-pupil and pupil-pupil discussion? Why was there so little diagnostic work? Why was the curriculum so dominated by formal mathematics schemes and so little influenced by children's spontaneous interests? Why did teachers with such an elaborate view of children's thinking cast their pupils into passive-receptive roles as learners or permit them to adopt such roles?

In responding to these questions on the basis of our interpretations of our observations, it seemed to us that a number of interacting factors stood between the teachers and the realization of the totality of their aims. Each factor pushed the teacher towards direct instruction and the practise of routine skills and away from enquiry methods and the encouragement of reflective thinking and general problem solving strategies. Each factor operated in reality in a way completely different from that implied in the exhortatory literature. The factors were; the programme of work, the children's skills and motives, the auditors of primary practice, the teachers' working conditions and the teachers' human status. These factors, we inferred, never operated independently of each other.

When it is suggested that teachers adopt a variety of teaching strategies — including enquiry methods in response to children's immediate or spontaneous interests — in order to meet the broad range of mathematical aims, including the attainment of basic routines and the higher order strategic skills of problem identification, choice of methods and the like, it is implied that the teacher's programme of work is sufficiently flexible to permit or encourage this, that children are (or can be) highly motivated to be so engaged, that these practices are what the immediate auditors (children, parents, other teachers, LEA officers) of children's and teachers' performances really want and that the teachers' working conditions are sufficiently well resourced for her to be able to respond productively. Deficiencies in any one of these factors, because of their interaction, is a threat to the whole venture. Every one of these was a limiting factor in the classrooms observed.

We have seen that the programmes of work observed were dominated by commercial mathematics schemes. The days of total spontaneity in the infant school have long since gone (in so far as they ever existed). Some programme is deemed necessary to secure the place of mathematics on a very crowded early years curriculum, to help attain continuity and progression and to act as a point of contact for parallel classes. These points were all made by the teachers who also noted that parents have a right to know what their children are doing and where it is all going. The teachers also suggested that parents can be informed on these matters by reference to a school programme. What the teachers did not mention is that their LEA (in common with most other LEAs) recommended that each school lay down guidelines for the mathematics content to be covered at various levels in the school (a practice endorsed by HMI). These suggestions constitute another force towards a programmatic approach to mathematics teaching.

Such demands and guidelines, of course, left the teachers with a great deal of autonomy. They only needed the energy, expertise, time and material resources necessary to exercise it.

In the event the teachers had exercised their autonomy by choosing a commercial mathematics scheme as the central element of their mathematics curriculum. Decisions on the specific scheme had been taken after school-wide discussions and in the light of advice from LEA advisers. Whilst the teachers were critical of the schemes chosen we inferred that the materials had some distinct attractions. In their absence the teachers would have had an enormous intellectual task to design a detailed structure of mathematics experience and an even bigger mechanical labour to produce the materials necessary to keep children working and to offer the kind of differentiation available in commercial schemes. These have a structure designed by experts and the publishers pay a lot of attention to producing copious quantities of very attractive materials. A teacher — given her cash and time resources — could not hope to compete with these productions.

Unfortunately as we have seen, these materials produce their own management problems. The work sheets necessitate a lot of task specific teaching because each exercise has production or recording features which go beyond the mathematical demands of the task. It is not always the case that these extra demands bamboozle the children. Indeed, in about half the cases observed the children knew what to do before they were taught. The problem was that the task demands were often unpredictable and to protect their children from the effects of failure or confusion the teachers were disposed to teach every task.

The structure of the schemes encourages a press for coverage — that is to say, a desire to complete the syllabus as defined by the scheme. If this is not aimed for then some of the purpose of the schemes (that is, continuity, collaboration) seems to be lost. If the press does not come from the teachers it certainly does from the children. The overt structure of the schemes expressed in terms of ostentatiously numbered cards or books had the children treating it as a race track despite their teachers' efforts to discourage this approach. The children appeared very much less interested in discussion work. Since the workcards were the main, if not the sole, source of assessed work we are tempted to conclude that Doyle's 'performance for grades' factor was operating here. But if it was, it was an arena very much chosen by the children. Even when the teachers wanted to pause for thought it seemed that the children could hardly wait to get back to their tables to 'get on' with their cards. Thus, partly because of the children's response but chiefly because of the demands of workcards, the teachers were under pressure to teach the elements of the scheme which pressure rapidly integrates to become simply, 'teach the scheme'. As we saw, this did not entirely squeeze out attention to spontaneity or to real problems brought in by the children — but it very nearly did and we might anticipate that increasing demands for standards would add to the pressure already exercised by the children. It seemed to us that whilst the schemes had obvious attractions they were a powerful mediator in constraining the quality of the pupils' learning experience in particular through the manner in which they were necessarily managed by the teachers and interpreted by the children.

The children exhibited all the impressive qualities attributed to them in the exhortatory literature on mathematics education. Taken individually they showed the capacity to reason, concentrate, make sense and deploy basic mathematics understanding. They did not limit the application of these qualities to the specific mathematics tasks assigned to them. Rather, they made sense of the totality of the situation they found themselves in. In short they adapted to the classroom, including the teacher and the mass of other children. They interpreted their tasks in the social circumstances in which they were set. At the general level, they could see no relevance of mathematics to their own or anyone else's life. Mathematics was, in their view, simply their work and they were happy to do it. They did not query the work. They did not tell the teachers they could already do it nor ask questions when they could not. They appeared to make no distinction between completion and comprehension. If they could complete a task, the work was done. If not, the teacher helped. In their well mannered, self controlled way they became another factor constraining the quality of their experience. In this respect they exhibited other characteristics common to children of this age range and well described by Piaget (1971) '... let there be no misunderstanding. Memory, passive obedience, imitation of the adult and the receptive factors in general are all as natural to the child as spontaneous activity' (pp 137–8).

The children were not cowed into the acceptance of their lot. On the contrary, the atmosphere in the classrooms could hardly have been more pleasant. They throughly enjoyed their work, a pleasure probably flowing from the sense of competence they must have felt as they made obvious progress through the work system. If ever it occurred to them to complain — in the fashion of an adolescent — about relevance, or to query — in the fashion of an intellectual — the approach to a problem, then it seemed likely that the good manners and caring-sharing attitudes they had been taught and which were essential to the smooth running of the classroom, would mitigate against their expressing a contrary opinion.

As we showed in chapter 2, the teachers were acutely aware of the dangers of the earnest desire to please exhibited by most children. On the one hand they recognized that they needed it to make humane management possible; on the other hand they saw the threat to the children's intellectual autonomy. In practice the conflict was resolved in favour of stable management rather than in favour of intellectual cut and thrust. Not the least reason for this, in our view, was that this was the children's preference too. Whilst the children were well mannered and well managed they appeared not to be equally accepting of all situations. They evinced much more concentration and less distraction when listening to instructions on how to do their cards than when taking part in discussion lessons. Their teachers recognized that engagement and control, whilst never a major problem, did become issues in the latter contexts. It seemed open discussions were less rewarding to both parties.

The children also exhibited little taste for, or ability at, sharing ideas on the kinds of tasks set in these classrooms. We saw in chapter 2 that the teachers had reservations about their children's facilities in this respect and had expressed more stringent reservations about their own capacities to arrange and make use of child-to-child conversations to enhance mathematical progress. Whether chaired by the teacher or left to their own devices we saw very little of this sort of exchange and only one which we thought productive. The children were adept at copying or telling each other what to do. They showed little sign of provoking each other's thinking. In this respect they were similar to children in many research studies. There is plenty of evidence to support the view that children can be effective tutors when adopting a direct instruction model of teaching. There is little evidence that they make much sense of each other's ideas in more subtle exchanges. Where such evidence exists, the exchanges have taken place on very carefully arranged — one might say contrived — tasks which have no curriculum substance. Also, effective exchanges, that is, exchanges which have been associated with intellectual advance, have flowed in these studies only between carefully chosen children, the choices having followed copious pre-experimental testing.

There is no question that children can learn from the social interaction with other children. A question mark, however, does hang over the conditions necessary for this to come about. Doise and Mugny (1984) concluded their thorough review of experimental studies of the social development of intellect as follows:

> ... our research clearly shows that progress will not arise from just any interaction or any task. For a child to learn from

> another child ... certain specific conditions must be fulfilled. Even at the risk of provoking the displeasure of advocates of various forms of 'spontaneity' in current educational debates, we feel that the application of our ideas to teaching in school will in no way reduce the importance of the adult's role. (pp 165 and 166)

It would seem that the teachers' views were entirely consistent with this opinion from the research. We are bound to conclude that the major reason we saw few child-child exchanges was because these are excruciatingly difficult to manage in a way that the teachers saw as fruitful and that in expressing these difficulties these teachers were asserting their honesty and insight rather than their incompetence. The teachers' views were consistent with those of leading researchers in the field.

Children's abilities and tastes and the substantial programmes of work were not, in our view, the only factors constraining the quality of mathematics work. There seemed to us to be several pressures which, although external to the classroom were very close to it. These comprised pressures from parents, headteachers, LEA officers and other teachers. The class teachers we studied claimed to ignore these external forces. However, they seemed to be subject to a welter of signals about what important people would count as a proper mathematics curriculum.

These signals, we inferred, were entirely consistent in their message and favoured work on routine skills. Some of the teachers mentioned that standardized tests of attainment were used by heads to monitor children's progress. These tests invariably assess routine productions. LEA guidelines emphasize contents rather than happenings. Parents liked to see (literally) what their children had done and to meet this demand necessitated the completion and recording of substantial amounts of work familiar to the parents' generation. Parents, we were told, often bought books and mathematics tests and exercises for their children (frequently at the children's request) and these books were mainly designed by authors claiming psychological eminence and educational expertise. They could be taken to form a high status source of definitions of appropriate mathematics experience for young children. It would be easy to assume that, despite all the rhetoric, what people really wanted as an early years mathematics curriculum was the rote acquisition of the so-called 'basic skills'.

All these pressures were taken by the teachers to be expressions of legitimate interest and we might predict that their potency — especially that of the parents — will increase as the power of parent governors is enhanced. We inferred that the teachers met these demands in all important respects and that they could hardly do otherwise. When they said that they ignored such forces it seemed clear that what they meant was that they ignored the demands for an extreme pace of progress through the curriculum in the cases of particular children whom they judged to be unready for the ambitious advances desired by some parents. That is to say that the teachers' resistance was genuine and enacted in the face of specific instances and in the interests, as the teachers saw them, of individual children. There was no resistance on matters of curriculum principle.

Most of the above pressures might be said to have their roots in tradition. Parents ask for what they were used to and were led to believe was valuable. The tests used in schools were conceived in terms of associationist learning theory and meant to assess children's immediate responses on aspects of rote memory. They strongly reflect a basic skills curriculum and this is precisely what primary classrooms were originally designed to provide. The classroom was the cheap building block in a mass processing system intended to transfer rudimentary knowledge and skill from teachers to children at minimal cost in time and money. Whilst the demands on the early years curriculum have expanded drastically, the basic structure of the system has remained the same. One adult in a small room with a few resources treats a large number of children.

Whilst the basic features of the educational delivery system have remained constant the curriculum is now expected to attend not only to the three Rs but to science, industry, the performing and expressive arts, the civics of a multicultural society, home-community links, special needs children and information technology.

The teachers in our study seemed very conscious of the amount of material they had to cover — not only in mathematics but in these other zones of the curriculum. When the authors of the Cockcroft Report exhort teachers to follow issue x or question y as these are spontaneously raised, they perhaps do not quite know what they are asking for. Our teachers allowed only slight scope for spontaneity and, seen in the context of the demands on the whole curriculum one can perhaps see why.

One extensively canvassed solution to the problem of fitting increased demands into finite time is to integrate issues across curriculum areas — to seize opportunities to teach or practice mathematics in art or science for example. Such a strategy has its attractions and was rated highly by some of the teachers, most notably Mrs. A. But we saw little of it in operation and in any event some of the teachers were deeply suspicious of it. If it is not to threaten the integrity of experience in both zones of the curriculum, such a strategy requires, at least, opportunistic skills of the very highest order. If these are not available this approach is like, to use Mrs. D.'s expression, '. . . mixing Scotch and water — it spoils two good things'. In these instances, integration can be utterly contrived and might even be counter-productive. Sorting animals by size or other qualities after a trip to the zoo for example, appeals to the adult but in the experience of our teachers children find this boring and odd.

Another solution to fitting the curriculum quart into the classroom pint pot has emerged in the form of the technology of microprocessors. Thus, instead of seeing these devices as an added curriculum burden they may be conceived as a valuable teaching aid. The reality in the classrooms we observed was that these materials were not available at all, or very strictly rationed or out of order. Even when available the software was, in the eyes of the teachers, of dubious quality and fit, in the main, for the exercise of basic skills rather than the enhancement of thinking. Of course, these artefacts show exciting promise but the delivery is likely to be expensive.

Added to the sheer width of a modern curriculum is the burgeoning aspiration to teach higher order — and hence more generalizable — skills such as learning to learn, problem analysis and the like. As we have noted frequently, these skills lie at the heart of the problem of modern mathematics teaching.

In the literature on mathematics education it has been implied that such skills are most likely to be acquired through application work in which children attempt to solve substantive problems — and preferably real-life problems — or attempt tasks which show a close resemblance to real issues outside the classroom. It appears that the characteristics of these problems should be that the solution is not immediately clear, that there may be distracting features and that alternative approaches might be fruitfully considered. The teachers were fully aware of this line of thought, endorsed it as an important aspect of mathematics experience and spelled out the dangers of an otherwise narrow treatment of mathematics. Yet we saw very little of this kind of work. When the teachers faced children with facsimile problems they were often successful in engaging their attention — and particularly so with money and shopping matters. But generally the concrete practical work involved in these and related instances swallowed up most of the allocated mathematics time: witness the daisy chain incident in chapter 2.

It is often assumed that the children can be relied upon as a good source of application questions and problems. We saw not a single instance of this. They seemed far too eager to get on with what they saw as their real work — the mathematics scheme. Of course they might have been socialized out of raising such matters but we do not find that a very convincing thought. As with the teachers, it seems to us that there is little that is mathematical — however broadly that is conceived — in the children's perceptions of their lives although it is always possible to contrive or impose some mathematical structure on most experiences. We took this to be what the teachers meant when they expressed their suspicion of the mathematics of children's every-day experience. They saw that it could be imposed but doubted the effects of such an imposition.

They also doubted the practicalities of responding to problems raised by children. In the teachers' experience children are not especially interested in each other's mathematically laden problems. To steal Mrs. D.'s phrase, it is not everyone who is interested in someone else's chicken coop.

Even if the interest were there, such problems, we inferred from the teachers' comments, might be useful to the academic leaders and talkers in a group of children but the teachers were dubious about what anyone else would get out of this work. It seemed that the teacher felt that such problems are not well structured to the needs of many children and that it was not easy to see how these problems could be developed in a structured way. Since these teachers seemed totally incapable of laziness or cynicism we judged their view to be realistic in the light of their experience. It seems implicit in the teachers' remarks that whilst they fully supported the notion of applications work and went some little way towards realizing it, they felt that for children of this age the idea had never been properly thought through in terms of identifying the sorts of skills that were expected or in designing the kinds of tasks and specifying the resources needed to foster these capacities. It seemed to us that, lacking the support of serious conceptual work in this area and, under the pressure of the forces described above, application work, for the most part, was not developed.

Also undeveloped was the teachers' diagnostic work essential to the assignment of tasks appropriate to their children's understanding. Most of the work we saw was very well within the grasp of the children but about one-third of tasks was beyond their comprehension. Some children were clearly under-stretched whilst others were out of their depth. The reason for this deficit appeared to us to be perfectly obvious: the teachers were completely outnumbered and they saw this as their most acute problem. Whilst the teachers did not recognize the constraining effects of the schemes or dismissed the effects of external opinion they were keenly aware of the sheer quantity of children in their charge and that the quantity effect was multiplied by their children's diversity. The teachers recognized that, ideally, each child needed detailed attention which they could not provide.

Extensive research has suggested that class size makes very little difference to the learning outcomes engendered by teachers (Slavin, 1984). But this research defines learning outcomes as scores on standardized tests of routine production. Such outcomes are not at issue. High scores on these tests can be readily attained through the techniques of direct instruction and these can be easily applied to the mass of children. What is at issue is the refinement of the children's understanding and this is hardly likely to proceed without appropriate diagnostic work.

Diagnostic work takes time, attention and training. Teachers — certainly of the generation studied — were not trained for this activity. Even when trained, our previous research (Bennett *et al*, 1984) showed that teachers find this excruciatingly difficult to carry out whilst at the same time as running their classes. Our research workers in this and previous projects have all found that establishing a child's grasp of concepts and skills requires enormous concentration and intellectual and social skills which inform the line of observation and questioning and sustain the child's interest and cooperation.

Earlier, we asked, 'Why do the teachers not conduct more diagnostic work?'. Our immediate and enduring answer is that it is extremely difficult to do this even without the responsibility of running the class and sustaining progress for a mass of children. Thus, whilst we can concede that class size has little influence on the acquisition of basic skills we cannot believe that class size is not a determinant of the attainment of higher order objectives.

The above constraint is a manifestation of the limits of human information processing. In order to make the best of their capacities, humans impose order on situations to make their interactions manageable. In recognition of momentary limitations of information processing it is considered that goals are often best achieved by careful, step by step planning and step by step performance. In this sense alone we can see the attractions and efficiency — in performance terms — of direct instruction.

Unfortunately these procedures are considered by some experts to be antithetical to the attainment of children's intellectual autonomy, particularly as evinced by the attainment of strategic skills. These higher order goals have been considered to necessitate activities on the part of the teacher which are much more reactive or responsive to pupils' ideas and views.

In the ideal case of one adult with one child and no important distractions this is not an impossible aspiration. Even the heretofore much maligned working class mum has been shown to be perfectly adept at sustaining profitable exchanges with her children. Indeed researchers (Tizard and Hughes, 1984) have suggested that such mums have a lot to teach teachers in this respect. But most of our teachers were mothers. Their children were highly successful and we have no reason to suspect that they had any deficits in maternal conversational skills. Yet these were not often in evidence in the classroom. The screamingly obvious difference between teachers and mothers is, of course, not of skill but of context. Even when they talked to one child, the teachers had to place a time limit on the conversation, could never forget the other thirty or so children in the class and could not totally decontextualize the conversations from the programme of work.

More typically, the demands of management had the teachers attempting to hold conversations or discussions with six or seven children whilst monitoring the work of the rest. We have seen the effects of these factors on their line of thought as they attempted to make sense of children's responses, maintain the flow and other social dimensions of the conversation and sustain its intellectual import. The result was the rather odd (in contrast to normal conversations), teacher-dominated exchanges. The teachers were well aware of the distracting influences but hardly noticed, at the time, the outcomes so limited in terms of their aspirations.

In all this the teachers were exhibiting perfectly well known human limitations. In short, some of the activities expected of teachers are perfectly reasonable in a situation with one or two children and with relaxed, relatively natural conditions. Knowing what we do about human thinking we might expect very little from classrooms which in all important respects could hardly be less well designed for these conversations.

Of course, from classroom conversations everyone may be expected to take away a little something. But the teachers were not disposed to be so blasé. Their very proper sense of responsibility and their awareness of the programme appeared to lead them to impose a pace and direction on exchanges. This approach as we have seen seemed to be exacerbated by the children's interpretation of the situation.

Overview

Teachers found themselves operating a crowded curriculum (not only in respect of mathematics) using tasks which demand a great deal of teaching merely so that their children could contend with the procedural features of their work. These factors presented a huge management load both in material and social terms.

In order to manage their day the teachers initiated children into certain social skills including helping, sharing, organizing, completing and cooperating. Children interpreted their tasks in this social setting and in the light of values and tastes brought by them into the classroom. Copious amounts of praise for work completed engaged the children's cooperation in management.

The upshot of these interacting forces of curriculum, management, instruction and cooperation was that, although there was some evidence of discussion and attempts at exploratory work, the predominant activity in the classrooms of these industrious and enlightened teachers was routine work happily connived at by children keen to cooperate and to force the pace of progress as they saw it.

Even when the curriculum was rendered relatively routine, there was still a high degree of unpredictability as the children interacted with the set tasks. Very large amounts of information about the children's intellectual responses were potentially available. The teachers could not conceivably respond to this information in its entirety. Their information selection rules operated — it seemed — mainly to sustain a good emotional tone in classroom relationships as indicated by the children's confidence and cooperation. To this end, teaching drove for work completion rather than comprehension. Completion was manifest, relatively certain and could be rewarded.

When teachers got off their routines the problems of sustaining the intellectually productive engagement of half a dozen children on a novel task, whilst at the same time necessarily monitoring the rest of the class, were acute. This problem is best interpreted as an information overload in a context in which rapid decisions are necessary. The teachers responded in typically human and limited ways.

The teachers' information processing interacted with the processes of curriculum, management and instruction — it seemed — to increase the quantities of routine behaviour perhaps best seen as the product of coping strategies.

Not surprisingly, tasks which took large amounts of teachers' time and undistracted attention (such as diagnosis, open-ended exploration and the like) were rare in these classrooms precisely because, it was inferréd, the teacher had neither of these commodities available to her.

The teachers might best be perceived as making a remarkably good best of a learning environment well suited to the inculcation of routine skills but which could hardly be less well designed for the conscious development of higher order thinking. In this light the problem of establishing and sustaining higher order skills in a mathematics curriculum is more daunting than it appears to be in previous analyses. By the same token, teachers' achievements may be seen to be that much more impressive.

Chapter 8

Working for Higher Order Skills

The teaching behaviours reported in chapter 4 are by no means unique to the teachers in this study. The preponderance of routine exercises, of pencil and paper work, of teacher-dominated conversation and the incidence of approximately 30 per cent of tasks overestimating the learners' attainments seem to be quite general. This picture is entirely consistent with our previous study of infant classrooms (Bennett *et al*, 1984), is clearly set out in HMI reports (HMI, 1978) and is revealed by major reviews of practice in the United States (Sirotnik, 1983). In these respects our teachers fit, at least on the surface, into a very general picture of common primary practice. In respect of the domination of teacher instigated routine pencil and paper work, the experience of mathematics seems to be a representative sub-sample of the general experience of primary education.

These well substantiated practices have not been happily received by scholars. Teachers have been accused of 'mediocrity' (*ibid*). They have been subjected to the well known harangues of John Holt (1964 and 1970) and the somewhat more mystical hectoring of Alexander (1984). In these eyes, practices in the primary classroom are notable only for their evidently alarming deficits. These critics imply that teachers lack commitment, will, effort, imagination, an appreciation of children's intellectual strengths and the industry to capitalize on them. It appears that teachers and those who train them have not sufficiently devoted themselves to the conceptual work necessary to the imaginative structure of young children's learning nor to the professional preparation crucial to the management of an appropriately challenging school day. In this interpretation, routine work is the consequence of the routine and complacent minds of teachers and teacher educators. What is needed, apparently, is more effort, more imagination, more awareness and a new language to conceptualize educational goals and processes (see Alexander (1984) for an extended development of this line of argument).

Our view is entirely at odds with the above interpretation. Whilst, as we have noted, at the surface level our teachers exhibited the commonly observed practices of the primary mathematics curriculum, our additional data (reported in chapters 2, 3, 5 and 6) show clearly the intellectual quality, insight, imagination and industry of the teachers as they struggled to deliver the curriculum under severe constraints. Due to the teachers' skills these limiting factors are not readily visible to the casual observer or to the zealot. These latter experts (like the sports fanatic who can hold any catch, score any goal or return any service) can teach all things to all children. Zealots squeal in a mixture of disappointment and incomprehension when professionals drop a chance, miskick a ball, fluff a return or — in this context — fail to turn each child into Renaissance man. We share the disappointment. We cannot share the incomprehension.

In digging beneath the surface of classroom practice we revealed the interacting factors constraining the teachers and to which they had to apply their energies on a relentless, moment-by-moment basis. The factors include a crowded curriculum (not only in mathematics) and an attendant push for coverage; the motives, skills and interpretive capacities of children (mainly, it appeared, directed towards making their lives busy but simple rather than complex); the low quality of conceptual support for the teachers (particularly in respect of the design of tasks in the mathematics schemes); and the sheer quantity of information processing necessary on the part of the teacher to sustain motivation and learning amongst thirty extremely diverse children. These factors were perceived to interact reciprocally with the teachers' management processes which they imposed to make their lives possible. We conclude that classrooms as presently conceived and resourced are simply not good places in which to expect the development of the sorts of higher order skills currently desired from a mathematics curriculum.

This is not to say that higher order skills were not exhibited in the classrooms studied. They were — and extensively so. The children organized their own comings and goings, resourced their own activities, monitored their own performances and were generally highly interpretive of the social situation in which they found themselves. All these skills of ordering and planning their classroom activities were expected of the children at age 5 and encouraged by the teachers' management techniques. The children quickly learned to use the classroom as a resource. What was missing was the application of these sorts of skills to mathematical problems or to the mathematics curriculum in general. Whilst the classrooms required higher order social skills because the children were held responsible rather than regimented, almost the reverse could be said of their specific experience in the mathematics curriculum. The teachers showed that with considerable effort and skill the classroom can be made a civilized and thought-demanding social environment. It did not follow — and does not follow — that it is a good environment for nurturing the sorts of academic skills that mathematics educators have in mind. How might the mathematical quality of this environment be improved?

Before developing the implications of our analysis we should point out that given the scale of the problem, ours was a very small study. The teachers and their classrooms were in no way representative of the vast variety of teachers and the conditions of teaching which prevail in infant schools. Indeed the teachers were chosen for their good reputations and this obviously marks them out as unrepresentative. Whilst we have shown that the teachers' classroom behaviours were very similar on the surface to those of teachers in many studies our critics could point out that there might be many different causes of these practices. This may be so. Some teachers might adopt the behaviours because they know no better, others because they are dictatorial by nature and yet others because they are lazy. Perhaps, it could be argued, only a few teachers are pressed into these practices by classroom processes. If this is the case then obviously the implication is that some teachers need training for wider horizons, others may need personality transplants and the lazy need encouraging — or sacking. Such interpretations of (and antidotes to) classroom practices are perfectly common in official reports. But when these problems are, as it were, cured, the newly enlightened and committed teachers will find themselves in exactly the position of the teachers studied in this project. The problem of classroom processes will remain.

The Way Ahead?

Whilst we have argued that classrooms and curriculums as they are currently conceived are unpromising environments for the enhancement of higher order problem-solving skills in young children, nothing we have said diminishes the attractions of these attainments as educational goals. If — as evidence suggests — pupils are adept at the rote reproduction of basic skills but cannot apply them to define and solve problems, then their hard won expertise is, to steal Whitehead's famous phrase, just so much 'inert knowledge'. In that sense pupils' basic skills are disconnected from the real world. The importance of breathing vitality into pupils' skills seems perfectly obvious and has not been lessened by our analysis.

It is worth observing that the problem of limitations in pupils' capacities to apply basic skills is not peculiar to mathematics. American studies suggest that children are not well able to use basic reading skills for effective learning and that problems with pupils' writing seem to reside in the qualities of thinking and organization rather than in the mechanics (Chipman, Segal and Glaser, 1986). As we have shown, the interacting forces of curriculum, accountability, instruction, information processing and management are fearsome constraints on the deployment of higher order thinking skills in classroom mathematics sessions. It would have been a surprise if they were not equally restricting in other areas of the curriculum. All that being said, it should be emphasized that none of these forces is a fixed quantity. There are no grounds for pessimism in our interpretation.

All teachers have considerable scope to influence the quality of experience in their classrooms. Our research is in essence about the limits of that scope. We feel we have studied teachers who have taken the potential of the modern classroom close to its limits. If we may be permitted an analogy, the primary classroom is the Ford Popular of educational provision. It is a very serviceable vehicle providing you do not want to go too far, too quickly and in lavish comfort. The vehicle can, by neglect, be reduced until it is a heap of junk. On the other hand, it can be tuned, customized, polished and otherwise nurtured to give long and efficient service. But it will never be a Rolls Royce nor a Formula 1 racing car. In the same image, the primary classroom can be an awful and neglected place or it can be a delightful place but it cannot deliver everything no matter how good the teacher. The classroom has design limitations. It seems to us that we have the choice of either making the very best of the classroom within its limitations (as, we would argue our teachers have done) or reconceptualizing some aspects of the learning environment.

Before developing some of the practical consequences of this line of argument we should recognize that there are at least two lobbies that could accuse us of being too late. In their view the problem is solved. Interestingly, these lobbies come from opposite ends of the spectrum of teaching ideologies. They are the direct instruction lobby and, for want of a better term, the 'inspired teacher' lobby.

Advocates of direct instruction can point out that the vast majority of people have been taught formally and have never, in their schooling, had the pleasure of opportunities to explore or experience enquiry methods. Yet a not insignificant fraction of the population exhibits the very capacities for higher order thinking which we wish to engender. There is merit in this argument. Additionally, there is some evidence that children construct new intellectual approaches to problems under rigorous conditions of drill (Resnick and Ford, 1981). We have recently shown that the effect of playing card games had a significant effect on 6-year-olds' facility with the concept of ten (Desforges *et al*, in preparation). Of course, card games are simply a way of disguising drill exercises. These facts cannot deny however that, whatever formal methods do for the intellectual capacities of high fliers, they appear to do little for the majority. The challenge of direct instruction is to research those circumstances in which drill can be productive for higher levels of invention and applications and from this work inform the design of a more vital curriculum for the mass of children.

The 'inspired teacher' lobby takes a different line. Set against formal, class teaching and all in favour of enquiry methods, these teachers recognize the problems of organization and management that their approach faces. But they see no problem that cannot — indeed has not — been solved by a mixture of inspiration and devotion. They claim to have broken through the very barriers that we suggest limited even our very able teachers. Their classrooms draw on the spontaneous skills and interests of children. They are in touch with the latest research on children's learning. The teachers have the capacity to monitor each individual child; they see when to intervene and when to leave alone. They advance learning with a task here, a comment there and they learn from every exchange. In this view, 'The skilled, intelligent infant teacher, who possesses also imagination and the will to go on learning, has the power to enhance the potential of thousands of small persons during a teaching lifetime' (Kernig, 1986).

It is extremely difficult to evaluate this sort of claim because we know of no objective research on such teachers. We have never seen this sort of practice despite earnestly searching for it over ten years. We have seen teachers operate like this for a few moments and we have seen classes operate like this for special events. But we have never seen this practice on a day-to-day basis. There are many in teacher education who claim to have conducted this sort of programme in the springtime of their teaching careers. We are suspicious of this sort of claim. Our suspicion is based on everything we know about the limits of human information processing. However, just as those who felt the four minute mile was impossible have been shown to be misguided, so we would like to have our scepticism removed. Teachers who feel they have broken the barriers of classroom constraints make the rest of us feel so inordinately guilty that they owe it to us to teach us their magic. They should immediately volunteer to be the subject of detailed, long-term, objective research so that every aspect of their practice can be rendered available for inspection and transmission to the rest of the profession. We share their aspiration. We need their skill.

We anticipate very little from this line. Inspired practitioners have been around a long time and whilst they have contributed a lot to the literature on aspirations, they have provided little to the accumulation of transmittable, practical knowledge on how to enhance teaching practice. They have dealt too often in vague slogans and images rather than in practicable guidance. Indeed, in implying — but never properly substantiating — that anything is possible given their special blend of insight and dedication, they have distracted attention from the serious analysis of classroom processes.

Whilst recognizing that lessons about improving teaching practice may be learned from both ends of the spectrum of teaching ideology we should caution that direct instruction offers well researched practices which are not, as yet, a solution and that inspired teachers offer solutions which are not, as yet, well researched. Whatever may be gleaned from the study of these practices we are convinced — in the light of our research — that the major barriers to the establishment of higher order skills in classrooms lay not in the practice of teaching but in the conditions of teaching. In that light our prescriptions for change are directed more at those who provide — materially and especially conceptually — for practice and only tangentially at those who execute it. That is to say that in so far as contemporary mathematics teaching practices in the infant school may be seen to fall short of expectations, the burden of responsibility, in the terms of our analysis, lies with the educational managers who — whether deliberately or by default — provide the crucial psychological parameters of the teaching environment to which teachers and children alike must adapt. It is on these same groups that the onus for change must lie.

Facing Some Immediate Problems

Clearly anything which adds impetus to the constraining forces we have identified is bound to make matters worse in the sense of driving teachers towards paying even more attention to overteaching basic skills and diverting attention away from the exploratory, open-ended and time consuming work likely to be necessary to the acquisition of problem solving capacities and applications work. Since completing the research reported here we have taken the opportunity of discussing our data and analyses with more than 1000 experienced first and infant school teachers on a variety of in-service courses. Almost without exception they have endorsed our view. Indeed, they have pointed out how acutely they feel the forces of accountability and curriculum overload. The general opinion is that both these aspects of classroom life are worsening at an alarming rate. The treatment of mathematics has to be seen in this general context.

Teachers have reported a relentless pressure to attend to an ever-increasing list of objectives. Their perceptions are not based on rumour. The strictures are manifest in torrents of HMI outpourings and a related — perhaps consequent — deluge of LEA 'discussion documents'. These require attention to science, technology, industry, multicultural issues, special needs children, parental involvement, aspects of community schooling and so on. It is not that any one of these matters is seen as inappropriate to the infant school curriculum but rather that the whole volume is seen as a curriculum kilolitre pouring into a classroom pint pot. In the light of our analysis we can predict that the only manageable response to this is a once-over-lightly treatment of basic facts and routines. There is simply no time for serious, thoughtful treatment.

It seems to us that the primary curriculum is long overdue for a clear-out of the sort manifest in the secondary sector GCSE aspiration. Rather than bolting on new elements at — it appears — a rate of one per week, it is evident that some drastic decisions have to be made about curriculum priorities in the light of judgments about what it is reasonable to attain rather than dreams about what is desirable. In addressing this problem, some experts (for example, Blenkin and Kelly, 1983) have suggested the adoption of a process curriculum. In brief, this entails the identification of significant learning and thinking processes with which we wish children to engage. A curriculum is then built around the design of opportunities to acquire and exercise such intellectual facilities.

There are problems with this approach not the least of which are (a) there is a lack of agreement on whether it is sensible to conceive of learning processes independent from content; (b) there is lack of agreement on what such processes might be; and (c) lists of skills can run to thousands of elements. In short, the idea might not provide the economies of time necessary to putting it into practice even if the general model were tenable.

We take a simpler view. In the absence of any convincing model of the curriculum as an alternative to that presently practised, we suggest that schools be expected to cover less ground and to use the time saved to cover it more thoroughly. Thorough here means going beyond the reproduction of skilled routines and into the intellectually challenging, practical application work. This proposal appears to run counter to HMI demands for a broad curriculum. We would suggest that breadth could be obtained by applying a small number of core concepts to a wide range of practical problems. Specifically with respect to mathematics, the concepts of measurement, money and four rules with small numbers would provide an ample set of issues to engage and develop children's applications skills up to the age of 11.

Obviously, we would not expect everyone to agree with this specification. The suggested approach immediately raises the question of how one decides what is reasonable to attain. For the moment we leave that as an open question but with the observation that it does not seem sensible for every individual class teacher — or indeed headteacher — to declare independence on the matter of key content. At the very least it looks like a matter to be decided at an LEA level. We would not want the question of 'who decides' however, to distract attention from the proposal that the primary curriculum as a whole is ripe for culling to create the time necessary for intelligent applications work in all aspects.

Limited time, we have argued, interacts with assessment or accountability atmospheres and practices to influence teachers' priorities. The teachers we have consulted have been made well aware of the persistent demands for an observable — if not measurable — educational product. Parents like to see — literally — what their children have done at school. They can see and appreciate pages of sums duly ticked. They cannot see esoteric conversations directed at problem solving. Adopting an approach to assessment similar to that of the parents, LEAs use standardized tests of mathematics attainment ostensibly to monitor curriculum coverage. Such tests assess the rote capacity to reproduce basic skills. They do not appraise comprehension, applications skills or general problem solving. Teachers are left in little doubt about what their customers and paymasters want. It is anticipated that these already stark signals will become sharper with the advent of increased parent power and efforts to assess teachers' productivity. Seen in this light, talk about teaching 'thinking skills' and 'fascination with mathematics' is just so much hot air. These activities are unlikely to appear in teachers' repertoires whilst the outcomes of their teaching are assessed in current terms.

Of course parents can be educated to appreciate educational outcomes more subtle than a page of sums. But parental education takes time. It is a never ending job. This should not be only the teachers' task. If LEA advisers believe their own rhetoric they should face up to its broader implications. They must strive to create the educational context which makes desired practices more likely. They should (after appropriate consultation) issue parents with guides to good practice consistent with their aspirations. In the same vein they should develop systems of pupil profiling to replace standardized testing and to reflect the priority of problem solving experience over routine exercises.

If teachers are to be assessed, then the systems used should also reflect the assessors' aspirations. Teaching behaviours which are thought to be associated with the engagement of pupils' thinking and problem solving capacities should have a very high priority. Of course these apparently simple and obvious suggestions hide some hideous problems. The first is that at the moment the desired thinking skills are so vaguely specified. The second is that 'appropriate teaching behaviours' are necessarily even less easy to define with any validity. However, the challenge of establishing suitable assessment techniques might aid all concerned in the concrete identification of these somewhat ephemeral processes. It is a challenge that can hardly be avoided since current procedures are so obviously counterproductive.

Nurturing Children's Mathematical Thinking: The Theorists' Responsibilities to the Teacher

We have argued that the development of problem solving skills requires time to explore, reflect and discuss and we have suggested that there is much that administrators and advisers must do to create both the time and the accountability atmosphere necessary to its proper use. Of course, there is much that teachers can also do in their classrooms to make the most of time available. Teachers can save time by suitable planning and organization aimed at reducing time lost in transitions between activities, in the general fetching and carrying necessary to any work and in the disruptions which seem endemic to primary classrooms. Time savings by teachers can be considerable. American research indicates that the least time-conscious teachers can lose up to 25 per cent of allocated teaching time (Anderson, 1984).

Grossed up over the school year this can amount to the same loss as taking more than a month off school.

Time saving is clearly attractive but there may be associated costs. Humans do not always take kindly to a relentless time and motion approach to their work. Young children can be over-organized and in our experience seem to wither and pale with the stress and fatigue that this causes. Additionally time spent 'fetching and carrying' is not always wasted. As introspection should show, it can serve an important winding up or incubation purpose allowing thoughts to be gathered or a mood to be set.

Teachers can also create time by making sure that routine skills are covered with maximum efficiency, that is, in minimum time. The system of mastery learning, for example, minimizes the time needed to learn basic facts and skills. But these savings are also not without cost. The rigorous application of mastery learning techniques would be expensive in terms of materials and appropriate testing devices.

Whatever savings are to be made, it would be foolish to squander them on extending current practices. Experience has taught us that repeated exercises involving the solution of standard problems does not automatically bring about the acquisition of problem solving capacities. Using basic skills in novel situations requires that pupils must learn how to analyze the problem, to search their memories for relevant routines, plan possible solutions and keep track of and evaluate progress as they work. Children clearly do not catch these abilities simply by being given large doses of problem solving experience.

Unfortunately, whilst the requirements for a general problem solving capacity have been well known for decades — see Polya (1957) for example — there has been little progress on establishing how these skills can be taught. Recent reviews of research and development in this field have shown that there is (a) a lack of evidence from normal classrooms working under normal conditions; (b) a lack of success with ordinary pupils under any circumstances; and (c) doubt in the minds of some experts as to whether such general skills exist (Chipman *et al*, 1986; Nickerson *et al*, 1985; Segal *et al*, 1986; Weinstein and Mayer, 1986).

In view of this research it seems to us that researchers have a duty to teachers to come clean and admit the profound difficulties they are facing — even in the relatively easy conditions of experimental studies — in enhancing the general quality of children's thinking by means of educational interventions. Rather than so doing, researchers seem to be prone to be over enthusiastic about their achievements given that they have precious little evidence as to the efficacy of their suggested approaches. Whilst researchers' no doubt have the best interests of children in mind their exhortations are often counterproductive since, when their suggested methods fail to work, the best teachers feel guilty and the worst have their cynicism with regard to theorists strongly reinforced.

We have been as guilty as any in endorsing practices in advance of adequate research and classroom development. For example, we have advised teachers that it is essential to conduct much more detailed diagnostic work if ever the needs of learners are to be identified let alone met. We have not been alone in making this suggestion (Bell *et al*, 1982; HMI, 1985) but we should have known better. In the light of our previous research (Bennett *et al*, 1984) in which the paucity of diagnostic exchanges in infant mathematics work was striking, we designed and conducted an in-depth, in-service course for teachers aimed at improving their diagnostic skills. Despite intensive teaching and monitoring and copious amounts of individual tuition, the participants on the course showed no progress in the acquisition of these skills. We had previously found it quite easy to train experienced teacher/researchers to conduct diagnostic interviewing with a good degree of success. The obvious difference between the teachers and the field workers was that the latter did not have the responsibility of running the class whilst conducting their interviews. The added burden of monitoring other children does not leave the interviewer with sufficient attention to concentrate on a productive line of observation or questioning.

There is an added problem with diagnostic interviewing in that whilst it might show what a child's problem with a concept is, it does not necessarily indicate what might be done with the problem. Once children have a powerful misconception the teacher has to deal not so much with skill acquisition as with skill reconstruction. Some approaches to re-teaching in this situation can add to the child's confusion. The business of re-constructing misconceptions has, to our knowledge, had little treatment from the research community. Teachers have been left to make what they can of it.

These heavy doses of reality do not diminish the attractions of diagnostic work in principle. Indeed there is an even greater attraction not previously articulated. It has been evident in our own work and that reported by others (Erlwanger, 1975; Fuson, 1982) that children can cope with and respond well to diagnostic exchanges. If these exchanges were ever to become a regular fact of classroom life we might anticipate that children, predicting that they would be asked about their thinking, might take greater responsibility for exploring solutions rather than devoting themselves to sussing out what the teacher wanted to know. We have sometimes observed that during a diagnostic interview children reconstruct their thinking to make progess in understanding the concept under exploration. When the teacher asks herself questions about the child and seeks to answer them with the child, learning opportunities for both parties can be exciting. In short, diagnostic work seems to be an essential pedagogic skill in an ideal world. Since, however, diagnostic work is extremely difficult to conduct in the classroom as presently conceived, it seems little more than a researcher's self indulgence to prescribe it.

If diagnostic work is ever going to be useful to teachers then it is evident that a great deal of work is necessary in all the following respects (a) the development of diagnostic procedures and guidelines; (b) the development of remediation techniques which enable children to reconstruct their understanding rather than merely learn a new habit; and (c) skills training for teachers.

Some work on these problems has been conducted in America. Case (1986) has developed principles for designing instruction directed at skill reconstruction. Seely-Brown (1983) has designed computer simulations to train student teachers to diagnose children's errors in basic mathematics procedures. But this work is neither well known nor well tested. Even if such work were generally available it would not necessarily transfer to normal classrooms unless the accountability atmosphere and time allocations had been altered in the manner indicated earlier.

At the same time as being hectored about matching and diagnostic interviewing, teachers are also having mountains of untested ideas on how to enhance their mathematics curriculum thrown at them by mathematics education enthusiasts. Most fashionable at the moment are the notions of teaching mathematics concepts through games and developing thinking skills through investigations.

Mathematics games have long had a presence in the early years' curriculum but they have not formed an integral part of that programme. Recently, however, they have been presented with increasing intensity to teachers as having a significant role to play in children's experience (Hewett, 1984; Hughes, 1986). Kamii (1985) goes further in suggesting that games can become the better part of the early years' mathematics curriculum. In Kamii's view, games set a context in which children can teach themselves, through their processes of invention, the key concepts and procedures in mathematics at the introductory stages. She argues, 'For centuries, educators have believed that young children learned arithmetic by having it taught . . . yet in reality children have been learning most of it in spite of instruction.'. She suggests a radical reconceptualization of the teacher's role, 'I am advocating eliminating all traditional instruction in first grade arithmetic and replacing it with two kinds of activity: situations in daily living (such as voting) and group games' (*ibid*, p xi). This is the stuff of a very attractive pedagogic revolution.

It turns out that Kamii's enthusiasm is based on very little data. After two years' experimentation with first grade classes she has produced no objective evidence to support the view that her games help children to improve their mathematical knowledge. In her judgment her game players (in contrast to a non-game playing control group) were, '... mentally more active and autonomous ...' (*ibid* p 275). Whilst we respect this professional judgment and endorse the constructivist philosophy which guides it, it is hardly the stuff to sustain a pedagogic revolution. Kamii, like so many enthusiasts, seems to be in far too much of a rush to present ideas to teachers in advance of the serious professional work necessary to establish the integrity of her approach in the school setting. Those who seek to sell pedagogic techniques to teachers ought also to design and purvey the indices of progress they aspire to foster. If they cannot do this to make objective contrasts between control and treatment groups in experimental settings, what price the teachers' chances of satisfying their customers that progress is being made?

The notion of developing thinking skills through investigations work is even less encumbered by evidence than the current passion for games. The term, 'mathematical investigation', '... has come to mean a type of work whose value lies more in the activity of solving the problem than in the solution itself. The aim is that children should start to think in a more independent way, rather than select "the right method" from a set of "standard methods" already learnt' (SEC, 1986, p 26). From 1988, mathematical investigations are to be incorporated into the new GCSE examinations. This move is seen to have implications for the primary school. 'The younger children are when they start this kind of work the better. If students do not meet these approaches until they are older, they tend to see them as a side-track and are eager to return to what they call "real maths" — in other words to the safety of the text book exercises' (*ibid*, p 29). In the event, topics for investigation work for infant schools are already being extensively pedalled on in-service courses for teachers.

Two observations are worth making. First, there is not a shred of systematic evidence that investigations actually enhance the intellectual processes they are intended to nurture. Secondly, investigations are high in risk, ambiguity and information processing demands as conceived here. We might therefore anticipate, in terms of our analysis, that classroom processes will operate to close down these risks and to convert open problems into predictable routines. We have already noted (Davis and McKnight, 1976) that investigations work has met this response in American high schools. Some research of our associate (Hine, in preparation) shows similar, powerful reduction procedures in English primary schools. Teachers in Hine's study, under the press of the forces we have described, have investigations converted into procedures and train their pupils to work through them in a systematic manner.

As presently put to teachers investigations work might survive as an afternoon option or a fun or marginal element of the curriculum — like plasticene on a wet playtime. However, if it is ever to be an important part of a programme of work, those who propose the notion to teachers must face up to the more serious business of (a) establishing beyond the level of the purveyor's natural enthusiasm that the idea actually works; (b) locating individual investigations into a structured programme facilitating intellectual growth over the longer term; and (c) changing the accountability system in which the tasks will inevitably be located and interpreted. Presently, all these problems are left to the teacher who can then be scoffed at for failure.

The final technique we wish to consider here is the use of child-to-child conversations in the development of mathematical thinking. This practice is to be distinguished from the teacher dominated conversations common in infant schools. The essence of the technique is that the children articulate ideas amongst themselves and learn to share and examine viewpoints. What is intended is to create opportunities for children to reflect on and spell out their ideas and to seek appropriate criteria for judging the ideas of others. The notion, like that of games, is not new. However, it appears to receive little serious attention in infant school practice — at least in so far as that is revealed in research. Whilst children are frequently seated in groups for mathematics they rarely work as a group. Typically each child works on his or her own topic. Such conversation as there is, is generally about materials or social factors (Bennett *et al*, 1984). Even when the children work on a common project, very little of the talk is mathematically laden and less still could be described as productive. The relative infrequency in practice of a familiar and highly rated idea might suggest certain difficulties in its enactment. We saw in chapter 2 that the teachers in this project, whilst not doubting their children's knowledge about and interest in mathematics topics, seriously questioned their own capacity to manage child-child exchanges in a productive way. This doubt was certainly reflected in their practice. There were very few managed incidences of this type of activity.

Intimations of practical difficulties notwithstanding, there are those who insist that the encouragement of child-child conversations on mathematics questions should be a key element of the curriculum (Cockcroft, 1982). Some (for example, Easley and Easley, 1983) go further and consider that such exchange should be at the heart of the early years' mathematics programme. Easley and Easley rest their enthusiasm in part on a constructivist philosophy of cognitive development and in part on four months' observation of the technique in practice in a Japanese elementary school. The Easleys report,

> The teachers at the Kitamaeo school seemed to place their highest priority on teaching children how to study mathematics largely through group work rather than in trying to teach them mathematical ideas and skills ... Because the teachers worked so hard and successfully at teaching this method, it seems there was only occasionally a need for them to teach concepts and skills directly. (p 4)

It appears that the normal technique of these teachers was to ask the children to work individually on a challenging problem and then have them discuss their answers and reasons with other members of their mixed ability group to try to achieve a consensus of opinion. A more able pupil acted as a group leader. The leader's role was not to supply the right answer for the less able but to provide encouragement and ask questions that would increase understanding.

Interestingly, despite their enthusiasm for this technique, Easley and Easley omit any record of the group interactions which so impressed them. To represent their ideas they report instances in which the teacher is chairperson. The quality of the Japanese first graders is illustrated in the following example in which the children were discussing the illustration:

Child 1: You can't add three and zero. There is no answer.
Teacher: Do you mean the answer is zero?
Child: Yes.
Teacher: Who agrees?
Child 2: You can't add three and zero and therefore your answer is three.
Teacher: You can't add or there is nothing to add — which is it?
Child 2: There is nothing to add.
Child 3: Since there are three here, even if there are none here, there are three.

What is immediately obvious about this illustration is that there are no pupil-pupil exchanges. The contributions seem to be led by the teacher's questions. In other words, the most devoted enthusiasts of learning through pupil-pupil exchanges seem somewhat short on exemplary material. This is not to say that the idea is unattractive in principle. On the contrary it is extremely attractive. But, once again, it seems to be being pushed at teachers without proper evaluation or — in this case — even without convincing exemplification.

In fact there is something very realistic in the Easley and Easley example in that the adult plays a vital role in sustaining the exchanges. This feature is entirely consistent with a broad range of research on the development of children's thinking in this age range. The impact of the adult is critical. In reviewing research on play for instance Bruner (1986) has identified four conditions that, '... strikingly increase the richness and length of play ...' one of which is, '... an adult nearby to provide support and response' (p 605). In a similar vein, Smith (1986) referring to the cognitive benefits of play, concluded that, 'The role of the adult is a vital one' (p 10). And, as we recorded earlier, Doise and Mugny (1984) concluded their review of research on the development of cognition through social interaction amongst young children with the contention that adults played a significant role in such exchanges — at the very least in determining tasks and organizing appropriate groupings. In short, whilst child-child exchanges do occur in classrooms and are, potentially at least, a powerful engine for cognitive progress, the adult clearly plays a significant steering role. The practical, pedagogic implication of this is that tasks which arrest children's interests and which have mathematical potential must be available in the classroom. More importantly, so must the teacher. Once again, we see the probability — but not the inevitability — of classroom resources limiting the enactment of an interesting idea. Teachers can so rarely be available.

In summarizing this commentary on theoreticians' current suggestions for the improvement of the early stages of mathematics education it can be noted that the separate activities of diagnosis, games, investigations and cooperative group work each requires, for its successful implementation, a lot of time for exploration and good quality adult support (that is the undistracted attention from someone who knows how the activity might develop). The tasks or exchanges need to be located in a coherent structure intended to enhance and reveal progress in mathematics over the longer term. What teachers actually get presented with is a rag bag of 'good activities'. Teachers are not presented with a coherent programme nor with the indices of progress necessary to satisfy themselves or their customers that worthwhile activities are taking place.

These omissions in prescription, and the data supporting prescription, are symptomatic of a much deeper problem. The problem is that mathematics education experts have failed to identify — and hence failed to operate in their studies — adequate conceptions of the structure of mathematics problem solving competence as it develops in this age range. In other words the ways of talking about 'problem solving', 'appreciating mathematics' or 'thinking mathematically' have been vague and remain so. The problems of defining these processes and appropriate criteria for monitoring their development are exceedingly difficult. Perhaps the major responsibility of theorists in this field is to admit the limitation of their basic work.

Rather than creating the aura that the only factor preventing the attainment of our aspirations in early years' mathematics teaching is the conservative practice of teachers, experts in the field should admit that they have yet to equip themselves — let alone the profession — with the conceptual tools adequate to the job.

Such an admission might draw more first school teachers into the kind of research work necessary. We have shown that teachers have a vast knowledge of children's responses to tasks. They are also very self-critical. Because they care about children it is very easy to make them feel guilty and — feeling guilty they withdraw in the face of self-confessed experts. In this way researchers throw away their best resource, leave teachers open to cheap political jibes and make teaching more difficult. As James said almost a century ago, 'The worst thing that can happen to a good teacher is to get a bad conscience ... our teachers are overworked already. Everyone who adds a jot or a tittle of unnessary weight to their burden is a foe of education. A bad conscience increases the weight of every other burden ...' (James, 1899, pp 13–14).

Overview

We set out on this investigation with the suspicion that the teachers' job is more complex than that assumed by those who advise them on how to teach mathematics. Put bluntly, we have found what teachers already know: teaching mathematics is very difficult. But we feel we have done more than that. We have shown that the job is more difficult than perhaps even teachers realize. We have demonstrated in detail how several constraining classroom forces operate in concert and how teachers' necessary management strategies exacerbate the problems of developing children's thinking.

The teachers in our study held elaborate views of children's learning and correlated views of appropriate teaching. They subscribed to the same aspirations for their children as those pronounced by mathematics experts. We showed that holding these aspirations and working with consummate industry were not enough to overcome the constraining factors of the classroom.

It was recognized that these constraints are not fixed factors. Experts working with just a few children or on a demonstration basis can do out-of-the-ordinary things. But teachers deliver the mathematics curriculum over long time-scales and as only part of an overladened programme of work. Under these circumstances, and in anticipation of an increased preoccupation with crudely conceived accountability technqiues, we anticipated that the quality of mathematics education will not improve. We suggested that current approaches to enhancing mathematics teaching do not take into account the complexity of the teachers' job.

Our analysis takes nothing away from the appeal of teaching higher order mathematics skills nor does it question children's inherent capacity to acquire such skills. Our work does not detract from the high quality efforts of developmental psychologists and mathematicians who study the growth of children's thinking, the structure of mathematics or the design of curriculum materials. We did point out, however, that such researchers often imply criticisms of teachers and that — in the terms of our analysis — these criticisms arise out of a naive view of teachers' working conditions. We feel we have shown just how unpromising these conditions are. The challenge is for researchers to collaborate with teachers and administrators to change the material and intellectual conditions.

We argued that the major responsibility for desired changes lies with LEA officers. HMIs and theorists who should work to provide teachers with (i) an assessment context which unavoidably rates processes over products; (ii) intellectual tools and curriculum conceptions appropriate to the rhetoric of aspirations; and (iii) the time and circumstances in which to operate desired techniques.

Bibliography

ACACE (1981) *Adult Numeracy Study: A Survey Conducted by Social Surveys (Gallup Poll) Ltd*, Leicester, ACACE.

ALEXANDER, R.J. (1984) *Primary Teaching*, London, Holt.

ANDERSON, L.W. (Ed) (1984) *Time and School Learning*, London, Croom Helm.

APU (1982) *Mathematical Development: Primary Survey Report No. 3*, London, HMSO.

ATWOOD, R. (1983) 'The interacting effects of task form and activity structure on students' task involvement and teacher evaluations', paper presented to the annual meeting of the American Educational Research Association, Montreal.

AUSUBEL, D.P. (1968) *Educational Psychology: A Cognitive View*, New York, Holt, Rinehart and Winston.

AUSUBEL, D.P., and ROBINSON, F.G. (1969) *School Learning: An Introduction to Educational Psychology*, New York, Holt, Rinehart and Winston.

BELL, A.W., COSTELLO, J., and KUCHEMANN, D.E. (1982) *A Review of Research in Mathematical Education: Research on Learning and Teaching*, Nottingham, Shell Centre for Mathematical Education.

BENNETT, S.N., DESFORGES, C., COCKBURN, A.D., and WILKINSON, B. (1984) *The Quality of Pupil Learning Experiences*, London, Lawrence Erlbaum Associates.

BEREITER, C., and SCARDAMALIA, M. (1977) *The Limits of Natural Development*, mimeo, Ontario Institute for Studies in Education.

BLENKIN, G.M., and KELLY, A.V. (Eds) (1983) *The Primary Curriculum in Action*, London, Harper.

BLOOM, B.S. (1976) *Human Characteristics and School Learning*, New York, McGraw Hill.

BROWN, B.B. (1968) *The Experimental Mind in Education*, New York, Harper and Row.

BROWN, G., and DESFORGES, C. (1979) *Piaget's Theory: A Psychological Critique*, London, Routledge and Kegan Paul.

BROWNELL, W.A. (1928) *The Development of Children's Number Ideas in The Primary Grades*, Chicago, University of Chicago Press.

BRUNER, J. (1986) 'On teaching thinking: An afterthought', in CHIPMAN, S.F. *et al* (Eds) *Thinking and Learning Skills: Research and Open Questions*, Hillsdale, NJ, Lawrence Erlbaum Associates.

BUXTON, L. (1981) *Do You Panic About Maths?*, London, Heinemann.

CALDERHEAD, J. (1985) 'Teachers' decision-making', in BENNETT, S.N. and DESFORGES, C. (Eds) *Recent Advances in Classroom Research*, Edinburgh, Scottish Academic Press.

CASE, R. (1986) 'A developmentally based approach to the problem of instructional design', in CHIPMAN, S.F. *et al.* (Eds) *Thinking and Learning Skills: Research and Open Questions*, Hillsdale, NJ, Lawrence Erlbaum Associates.

CHIPMAN, S.F., SEGAL, J.W., and GLASER, R. (Eds) (1986) *Thinking and Learning Skills: Research and Open Questions*, Hillsdale, NJ, Lawrence Erlbaum Associates.

CLARK, C.M., and PETERSON, P.L. (1986) 'Teachers' thought processes', in WITTROCK, M. (Ed) *Handbook of Research on Teaching*, New York, Macmillan.

COCKBURN, A. (1986) 'An empirical study of classroom processes in infant mathematics education', unpublished Ph.D. thesis, University of East Anglia.

COCKCROFT, W.H. (1982) *Mathematics Counts*, London, HMSO.

CORNBLETH, C., and KORTH, W. (1983) 'Doing the work: Teacher perspectives and the meanings of responsibility,' paper presented to the annual meeting of the American Educational Research Association, Montreal

DAVIS, R.B., and MCKNIGHT, C. (1976) 'Conceptual, heuristic and s-algorithmic approaches in mathematics teaching', *Journal of Children's Mathematical Behaviour*, 1, pp. 271–86.

DEPARTMENT OF EDUCATION AND SCIENCE (1967) *Children and Their Primary Schools* (Plowden Report), London, HMSO.

DESFORGES, A., and DESFORGES, C. (1980) 'Number based strategies of sharing in young children', *Educational Studies*, 6, (2) pp. 97–109.

DESFORGES, C. (1978) 'Professional competence and craft performance in the reception class', paper presented to the British Psychological Society (Education Section) annual conference.

DESFORGES, C. (1985) 'Matching tasks to children', in BENNETT, S.N., and DESFORGES, C. (Eds) *Recent Advances in Classroom Research*, Edinburgh, Scottish Academic Press.

DESFORGES, C. *et al.* (in preparation) 'The effect of playing maths games on children's understanding of the concept of ten'.

DESFORGES, C., BENNETT, N., and COCKBURN, A.D., (1985) 'Understanding the quality of pupil learning experience', in ENTWISTLE, N.J. (Ed) *New Directions in Educational Psychology*, Lewes, Falmer Press.

DEWEY, J. (1916) *Democracy and Education*, New York, Macmillan.

DOISE, W., and MUGNY, G. (1984) *The Social Development of the Intellect*, Oxford, Pergamon.

DONALDSON, M. (1978) *Children's Minds*, London, Fontana.

DOYLE, W. (1983) 'Academic work', *Review of Educational Research*, 53, pp. 159–200.

DOYLE, W. (1986) 'Classroom organization and management', in WITTROCK, M. (Ed) *Handbook of Research on Teaching*, New York, Macmillan.

DUFFY, G.G., and ROEHLER, L.R. (1985) 'Conceptualizing instructional explanation', paper presented to the annual meeting of the American Educational Research Association, Chicago

EASLEY, J., and EASLEY, E. (1983) 'What's there to talk about in arithmetic?,' paper presented to the annual meeting of the American Educational Research Association, Montreal

EISNER, E. (1983) 'Can educational research inform educational practice?,' paper presented to the annual meeting of American Educational Research Association, Montreal

EMMER, E., EVERTSON, C., and ANDERSON, L. (1980) 'Effective classroom management at the beginning of the school year', *Elementary School Journal*, 80, (5), pp. 219–31.

ERLWANGER, S. (1975) *The Observation-Interview Method and some Case Studies*, mimeo, University of Illinois.

FEIMAN-NEMSER, S., and LODEN, R.E. (1986) 'The cultures of teaching', in WITTROCK, M.C. (Ed) *Handbook of Research on Teaching*, New York, Macmillan.

FEY, J.T. (1979) 'Mathematics teaching today: Perspectives from three national surveys', *Arithmetic Teacher*, 27, pp. 10–14.

FUSON, K.C. (1982) 'Analysis of the counting-on procedure', in CARPENTER, T.P. *et al.* (Eds) *Addition and Subtraction: A Cognitive Perspective*, Hillsdale, NJ, Lawrence Erlbaum Associates.

GELMAN, R. (1977) 'How young children reason about small numbers', in CASTELLAN, N.J., PISONI, D.B., and POTTS, G.R. (Eds) *Cognitive Theory*, Vol 2, Hillsdale, NJ, Lawrence Erlbaum Associates.

GELMAN, R., and GALLISTEL, C.R. (1978) *The Child's Understanding of Number*, Cambridge, MA, Harvard University Press.

GLASER, R., PELLEGRINO, J.W., and LESGOLD, A.M. (1977) 'Some directions for a cognitive psychology of instruction', in LESGOLD, A.M. *et al.* (Eds) *Cognitive Psychology and Instruction*, New York, Plenum.

GOOD, T. (1979) 'Teacher effectiveness', *The Elementary School Journal of Teacher Education*, 30, pp. 52–64.

HAYDEN, R.W. (1983) 'An historical view of the "new mathematics"', paper presented to the annual meeting of the American Educational Research Association, Montreal

HEWETT, I.V. (1984) *Number Games for Young Children*, Nantwich, Shira.

HINE, M. (in process) 'An exploration of children's task definitions in language and mathematics'.

HMI (1985) *Mathematics From 5–16: Curriculum Matters (3)*, London, HMSO.

HMI (1978) *Primary Education in England: A Survey by HM Inspectors of Schools*, London, HMSO.

HOLT, J. (1964) *How Children Fail*, London, Penguin.

HOLT, J. (1970) *The Underachieving School*, New York, Pitman.

HUGHES, M. (1986) *Children and Number: Difficulties in Learning Mathematics*, Oxford, Blackwell.

JACKSON, P. (1968) *Life in Classrooms*, New York, Holt.

JAMES, W. (1899) *Talks to Teachers*, London, Longmans.

JORGENSEN, G.W. (1977) 'Relationship of classroom behaviour to the accuracy of the match between material difficulty and student ability', *Journal of Educational Psychology*, 69, 1, pp. 24–32.

KAMII, C. (1985) *Young Children Re-invent Arithmetic*, New York Teachers College Press.

KERNIG, W. (1986) 'The infant looks at the world', in DAVIS, R. (Ed) *The Infant School: Past, Present and Future*, London, Institute of Education.

KUHN, D. (1979) 'The application of Piaget's theory of cognitive development to education', *Harvard Educational Review*, 49, 3, pp. 340–360.

LUNDGREN, U.P. (1977) *Model Analysis of Pedagogical Processes*, Stockholm, Stockholm Institute of Education.

MACKAY, R. (1978) 'How teachers know: A case of epistemological conflict', *Sociology in Education*, 51, pp. 177–87.

MEHAN, H. (1974) 'Accomplishing classroom lessons', in CICOUREL, A.V. *et al.* (Eds) *Language Use and School Performance*, New York, Academic Press.

MORINE-DERSHIMER, G. (1983) 'Instructional strategy and the "creation" of classroom status', *American Educational Research Journal*, 20, pp. 645–661.

NICKERSON, R.S., PERKINS, D.N., and SMITH, E.E. (1985) *The Teaching of Thinking*, Hillsdale, NJ, Lawrence Erlbaum Associates.

NUFFIELD (1967) *Mathematics Teaching Project: Mathematics Begins*, London, Chambers.

PAGE, D.A. (1983) 'Some problems and some proposed solutions or here is one horse to ride in the merry go round's next trip', paper presented to the annual meeting of the American Educational Research Association, Montreal.

PIAGET, J. (1971) *Science of Education and the Psychology of the Child*, London, Longmans.

POLYA, G. (1957) *How to Solve It*, New York, Anchor.

RESNICK, L.B., and FORD, A. (1981) *The Psychology of Mathematics for Instruction*, Hillsdale, NJ, Lawrence Erlbaum Associates.

ROMBERG, T.A., and CARPENTER, T.P. (1986) 'Resesarch on teaching and learning mathematics', in WITTROCK, M. (Ed) *Handbook of Research on Teaching*, New York, Macmillan.

ROSENSHINE, B., and STEVENS, R. (1986) 'Teaching functions', in WITTROCK, M.C. (Ed) *Handbook of Research on Teaching*, New York, Macmillan.

SCHOOLS EXAMINATION COUNCIL (1986) *Mathematics: GCSE — A Guide for Teachers*, Milton Keynes, Open University Press for the Schools Examination Council.

SEELY-BROWN, J. (1983) 'Learning by doing revisited for electronic learning environments', in WHITE. M.A. (Ed) *The Future of Electronic Learning*, Hillsdale, NJ, Lawrence Erlbaum Associates.

SEGAL, J.W., CHIPMAN, S.F., and GLASER, R. (Eds) (1986) *Thinking and Learning Skills: Relating Instruction to Research*, Hillsdale, NJ, Lawrence Erlbaum Associates.

SIEBER, R.T. (1981) 'Socialization implications of school discipline, or how first graders are taught to "listen"', in SIEBER, R.T. and GORDON, A.J. (Eds) *Children and Their Organizations: Investigations in American Culture*, Boston, MA, Hall.

SIROTNIK, K.A. (1983) 'What you see is what you get — consistency, persistency and mediocrity in classrooms', in *Harvard Educational Review*, *53*, pp. 16–31.

SLAVIN, R. (1984) 'Meta-analysis in education: how has it been used?', in *Educational Researcher, 13*, (8) pp. 6–15.

SMITH, L.M., and GEOFFREY, W. (1968) *The Complexities of an Urban Classroom*, New York, Holt, Rinehart and Winston.

SMITH, P.K. (Ed) (1986) *Children's Play: Research, Development and Practical Applications*, New York, Gordon and Breach.

STEPHENS, W.M. and ROMBERG, T.A. (1985) 'Reconceptualising the role of the mathematics teacher', Paper presented to the annual conference of AERA.

SUYDAM, M., and OSBORNE, A. (1977) *The Status of Pre-college Science, Mathematics, and Social Studies Education: 1955–1975. (Vol. 2: Mathematics Education)*, Columbus, The Ohio State University Centre for Science and Mathematics Education.

TITCHENER, E.B. (1912) 'The schema of instrospection', in *American Journal of Psychology, 23*, pp. 485–508.

TIZARD, B., and HUGHES, M. (1984) *Young Children Learning*, London, Fontana.

TRAFTON, P.R. (1979) 'Toward a better-balanced curriculum', in *Arithmetic Teacher, 26*, 2, p. 59.

WARD, G., and ROWE, J. (1985) 'Teachers' praise: some unwanted side effects', in *Society for Extension of Educational Knowledge, 1*, pp. 2–4.

WARD, M. (1979) *Mathematics for the The Ten Year Old*, Bungay, Suffolk, Schools Council, SCWP No. 61.

WASON, P.C., and JOHNSON-LAIRD, P.M. (1972) *Psychology of Reasoning: Content and Structure*, London, Batsford.

WEINSTEIN, C.F., and MAYER, R.F. (1986) 'The teaching of learning strategies', in WITTROCK, M. (Ed) *Handbook of Research on Teaching*, New York, Macmillan.

WEISS, I. (1978) 'Report of the 1977 national survey of science, mathematics, and social studies education', Research Triangle Park, N.C., Research Triangle Institute.

WELCH, W. (1978) 'Science education in Urbanville: a case study', in STAKE, R., and EASLEY, J. (Eds) *Case Studies in Science Education*, Urbana, IL, University of Illinois.

WOODS, P. (1976) 'Having a laugh: an antidote to schooling', in HAMMERSLEY, M., and WOODS, P. (Ed) *The Process of Schooling: A Sociological Reader*, London, Routledge and Kegan Paul.

YINGER, R.J. (1980) 'A study of teacher planning', in *Elementary School Journal, 80*, pp. 107–127.

Index